Introduction to Stochastic Processes

Introduction to Stochastic Processes

Paul G. Hoel
Sidney C. Port
Charles J. Stone
University of California, Los Angeles

Long Grove, Illinois

For information about this book, contact:
  Waveland Press, Inc.
  4180 IL Route 83, Suite 101
  Long Grove, IL 60047-9580
  (847) 634-0081
  info@waveland.com
  www.waveland.com

Copyright © 1972 by Houghton Mifflin Company. Reprinted by arrangement with Houghton Mifflin Company, Boston.
Reissued 1987 by Waveland Press, Inc.

10-digit ISBN 0-88133-267-4
13-digit ISBN 978-0-88133-267-4

All rights reserved. No part of this book may be reproduced, stored in a retrieval system, or transmitted in any form or by any means without permission in writing from the publisher.

Printed in the United States of America

21 20

Preface

In recent years there has been an ever increasing interest in the study of systems which vary in time in a random manner. Mathematical models of such systems are known as stochastic processes. In this book we present an elementary account of some of the important topics in the theory of such processes. We have tried to select topics that are conceptually interesting and that have found fruitful application in various branches of science and technology.

A *stochastic process* can be defined quite generally as any collection of random variables $X(t)$, $t \in T$, defined on a common probability space, where T is a subset of $(-\infty, \infty)$ and is thought of as the time parameter set. The process is called a *continuous parameter process* if T is an interval having positive length and a *discrete parameter process* if T is a subset of the integers. If the random variables $X(t)$ all take on values from the fixed set $\mathscr{S}$, then $\mathscr{S}$ is called the *state space* of the process.

Many stochastic processes of theoretical and applied interest possess the property that, given the present state of the process, the past history does not affect conditional probabilities of events defined in terms of the future. Such processes are called *Markov processes*. In Chapters 1 and 2 we study *Markov chains*, which are discrete parameter Markov processes whose state space is finite or countably infinite. In Chapter 3 we study the corresponding continuous parameter processes, with the "Poisson process" as a special case.

In Chapters 4–6 we discuss continuous parameter processes whose state space is typically the real line. In Chapter 4 we introduce *Gaussian processes*, which are characterized by the property that every linear combination involving a finite number of the random variables $X(t)$, $t \in T$, is normally distributed. As an important special case, we discuss the *Wiener process*, which arises as a mathematical model for the physical phenomenon known as "Brownian motion."

In Chapter 5 we discuss integration and differentiation of stochastic processes. There we also use the Wiener process to give a mathematical model for "white noise."

In Chapter 6 we discuss solutions to nonhomogeneous ordinary differential equations having constant coefficients whose right-hand side is either a stochastic process or white noise. We also discuss estimation problems involving stochastic processes, and briefly consider the "spectral distribution" of a process.

This text has been designed for a one-semester course in stochastic processes. Written in close conjunction with *Introduction to Probability Theory*, the first volume of our three-volume series, it assumes that the student is acquainted with the material covered in a one-semester course in probability for which elementary calculus is a prerequisite.

Some of the proofs in Chapters 1 and 2 are somewhat more difficult than the rest of the text, and they appear in appendices to these chapters. These proofs and the starred material in Section 2.6 probably should be omitted or discussed only briefly in an elementary course.

An instructor using this text in a one-quarter course will probably not have time to cover the entire text. He may wish to cover the first three chapters thoroughly and the remainder as time permits, perhaps discussing those topics in the last three chapters that involve the Wiener process. Another option, however, is to emphasize continuous parameter processes by omitting or skimming Chapters 1 and 2 and concentrating on Chapters 3–6. (For example, the instructor could skip Sections 1.6.1, 1.6.2, 1.9, 2.2.2, 2.5.1, 2.6.1, and 2.8.) With some aid from the instructor, the student should be able to read Chapter 3 without having studied the first two chapters thoroughly. Chapters 4–6 are independent of the first two chapters and depend on Chapter 3 only in minor ways, mainly in that the Poisson process introduced in Chapter 3 is used in examples in the later chapters. The properties of the Poisson process that are needed later are summarized in Chapter 4 and can be regarded as axioms for the Poisson process.

The authors wish to thank the UCLA students who tolerated preliminary versions of this text and whose comments resulted in numerous improvements. Mr. Luis Gorostiza obtained the answers to the exercises and also made many suggestions that resulted in significant improvements. Finally, we wish to thank Mrs. Ruth Goldstein for her excellent typing.

Table of Contents

1 Markov Chains — 1
- 1.1 Markov chains having two states — 2
- 1.2 Transition function and initial distribution — 5
- 1.3 Examples — 6
- 1.4 Computations with transition functions — 12
  - 1.4.1 Hitting times — 14
  - 1.4.2 Transition matrix — 16
- 1.5 Transient and recurrent states — 17
- 1.6 Decomposition of the state space — 21
  - 1.6.1 Absorption probabilities — 25
  - 1.6.2 Martingales — 27
- 1.7 Birth and death chains — 29
- 1.8 Branching and queuing chains — 33
  - 1.8.1 Branching chain — 34
  - 1.8.2 Queuing chain — 36
- Appendix
- 1.9 Proof of results for the branching and queuing chains — 36
  - 1.9.1 Branching chain — 38
  - 1.9.2 Queuing chain — 39

2 Stationary Distributions of a Markov Chain — 47
- 2.1 Elementary properties of stationary distributions — 47
- 2.2 Examples — 49
  - 2.2.1 Birth and death chain — 50
  - 2.2.2 Particles in a box — 53
- 2.3 Average number of visits to a recurrent state — 56
- 2.4 Null recurrent and positive recurrent states — 60
- 2.5 Existence and uniqueness of stationary distributions — 63
  - 2.5.1 Reducible chains — 67
- 2.6 Queuing chain — 69
  - 2.6.1 Proof — 70

	2.7	Convergence to the stationary distribution	72
		Appendix	
	2.8	Proof of convergence	75
		2.8.1 Periodic case	77
		2.8.2 A result from number theory	79
3	**Markov Pure Jump Processes**	84
	3.1	Construction of jump processes	84
	3.2	Birth and death processes	89
		3.2.1 Two-state birth and death process	92
		3.2.2 Poisson process	94
		3.2.3 Pure birth process	98
		3.2.4 Infinite server queue	99
	3.3	Properties of a Markov pure jump process	102
		3.3.1 Applications to birth and death processes	104
4	**Second Order Processes**	111
	4.1	Mean and covariance functions	111
	4.2	Gaussian processes	119
	4.3	The Wiener process	122
5	**Continuity, Integration, and Differentiation of Second Order Processes**	128
	5.1	Continuity assumptions	128
		5.1.1 Continuity of the mean and covariance functions	128
		5.1.2 Continuity of the sample functions	130
	5.2	Integration	132
	5.3	Differentiation	135
	5.4	White noise	141
6	**Stochastic Differential Equations, Estimation Theory, and Spectral Distributions**	152
	6.1	First order differential equations	154
	6.2	Differential equations of order n	159
		6.2.1 The case $n = 2$	166
	6.3	Estimation theory	170
		6.3.1 General principles of estimation	173
		6.3.2 Some examples of optimal prediction	174
	6.4	Spectral distribution	177
	Answers to Exercises	190
	Glossary of Notation	199
	Index	201

1 Markov Chains

Consider a system that can be in any one of a finite or countably infinite number of states. Let $\mathscr{S}$ denote this set of states. We can assume that $\mathscr{S}$ is a subset of the integers. The set $\mathscr{S}$ is called the *state space* of the system. Let the system be observed at the discrete moments of time $n = 0, 1, 2, \ldots$, and let X_n denote the state of the system at time n.

Since we are interested in non-deterministic systems, we think of X_n, $n \geq 0$, as random variables defined on a common probability space. Little can be said about such random variables unless some additional structure is imposed upon them.

The simplest possible structure is that of independent random variables. This would be a good model for such systems as repeated experiments in which future states of the system are independent of past and present states. In most systems that arise in practice, however, past and present states of the system influence the future states even if they do not uniquely determine them.

Many systems have the property that given the present state, the past states have no influence on the future. This property is called the *Markov property*, and systems having this property are called *Markov chains*. The Markov property is defined precisely by the requirement that

(1) $\quad P(X_{n+1} = x_{n+1} \mid X_0 = x_0, \ldots, X_n = x_n) = P(X_{n+1} = x_{n+1} \mid X_n = x_n)$

for every choice of the nonnegative integer n and the numbers $x_0, \ldots, x_{n+1}$, each in $\mathscr{S}$. The conditional probabilities $P(X_{n+1} = y \mid X_n = x)$ are called the *transition probabilities* of the chain. In this book we will study Markov chains having *stationary* transition probabilities, i.e., those such that $P(X_{n+1} = y \mid X_n = x)$ is independent of n. From now on, when we say that X_n, $n \geq 0$, forms a Markov chain, we mean that these random variables satisfy the Markov property and have stationary transition probabilities.

The study of such Markov chains is worthwhile from two viewpoints. First, they have a rich theory, much of which can be presented at an elementary level. Secondly, there are a large number of systems arising in practice that can be modeled by Markov chains, so the subject has many useful applications.

In order to help motivate the general results that will be discussed later, we begin by considering Markov chains having only two states.

1.1. Markov chains having two states

For an example of a Markov chain having two states, consider a machine that at the start of any particular day is either broken down or in operating condition. Assume that if the machine is broken down at the start of the nth day, the probability is p that it will be successfully repaired and in operating condition at the start of the $(n + 1)$th day. Assume also that if the machine is in operating condition at the start of the nth day, the probability is q that it will have a failure causing it to be broken down at the start of the $(n + 1)$th day. Finally, let $\pi_0(0)$ denote the probability that the machine is broken down initially, i.e., at the start of the 0th day.

Let the state 0 correspond to the machine being broken down and let the state 1 correspond to the machine being in operating condition. Let X_n be the random variable denoting the state of the machine at time n. According to the above description

$$P(X_{n+1} = 1 \mid X_n = 0) = p,$$
$$P(X_{n+1} = 0 \mid X_n = 1) = q,$$

and

$$P(X_0 = 0) = \pi_0(0).$$

Since there are only two states, 0 and 1, it follows immediately that

$$P(X_{n+1} = 0 \mid X_n = 0) = 1 - p,$$
$$P(X_{n+1} = 1 \mid X_n = 1) = 1 - q,$$

and that the probability $\pi_0(1)$ of being initially in state 1 is given by

$$\pi_0(1) = P(X_0 = 1) = 1 - \pi_0(0).$$

From this information, we can easily compute $P(X_n = 0)$ and $P(X_n = 1)$. We observe that

$$\begin{aligned}P(X_{n+1} = 0) &= P(X_n = 0 \text{ and } X_{n+1} = 0) + P(X_n = 1 \text{ and } X_{n+1} = 0)\\&= P(X_n = 0)P(X_{n+1} = 0 \mid X_n = 0)\\&\quad + P(X_n = 1)P(X_{n+1} = 0 \mid X_n = 1)\\&= (1 - p)P(X_n = 0) + qP(X_n = 1)\\&= (1 - p)P(X_n = 0) + q(1 - P(X_n = 0))\\&= (1 - p - q)P(X_n = 0) + q.\end{aligned}$$

1.1. Markov chains having two states

Now $P(X_0 = 0) = \pi_0(0)$, so
$$P(X_1 = 0) = (1 - p - q)\pi_0(0) + q$$
and
$$P(X_2 = 0) = (1 - p - q)P(X_1 = 0) + q$$
$$= (1 - p - q)^2 \pi_0(0) + q[1 + (1 - p - q)].$$

It is easily seen by repeating this procedure n times that

(2) $$P(X_n = 0) = (1 - p - q)^n \pi_0(0) + q \sum_{j=0}^{n-1} (1 - p - q)^j.$$

In the trivial case $p = q = 0$, it is clear that for all n
$$P(X_n = 0) = \pi_0(0) \quad \text{and} \quad P(X_n = 1) = \pi_0(1).$$

Suppose now that $p + q > 0$. Then by the formula for the sum of a finite geometric progression,
$$\sum_{j=0}^{n-1} (1 - p - q)^j = \frac{1 - (1 - p - q)^n}{p + q}.$$

We conclude from (2) that

(3) $$P(X_n = 0) = \frac{q}{p + q} + (1 - p - q)^n \left(\pi_0(0) - \frac{q}{p + q}\right),$$

and consequently that

(4) $$P(X_n = 1) = \frac{p}{p + q} + (1 - p - q)^n \left(\pi_0(1) - \frac{p}{p + q}\right).$$

Suppose that p and q are neither both equal to zero nor both equal to 1. Then $0 < p + q < 2$, which implies that $|1 - p - q| < 1$. In this case we can let $n \to \infty$ in (3) and (4) and conclude that

(5) $$\lim_{n \to \infty} P(X_n = 0) = \frac{q}{p + q} \quad \text{and} \quad \lim_{n \to \infty} P(X_n = 1) = \frac{p}{p + q}.$$

We can also obtain the probabilities $q/(p + q)$ and $p/(p + q)$ by a different approach. Suppose we want to choose $\pi_0(0)$ and $\pi_0(1)$ so that $P(X_n = 0)$ and $P(X_n = 1)$ are independent of n. It is clear from (3) and (4) that to do this we should choose
$$\pi_0(0) = \frac{q}{p + q} \quad \text{and} \quad \pi_0(1) = \frac{p}{p + q}.$$

Thus we see that if X_n, $n \geq 0$, starts out with the initial distribution
$$P(X_0 = 0) = \frac{q}{p + q} \quad \text{and} \quad P(X_0 = 1) = \frac{p}{p + q},$$

then for all n

$$P(X_n = 0) = \frac{q}{p+q} \quad \text{and} \quad P(X_n = 1) = \frac{p}{p+q}.$$

The description of the machine is vague because it does not really say whether X_n, $n \geq 0$, can be assumed to satisfy the Markov property. Let us suppose, however, that the Markov property does hold. We can use this added information to compute the joint distribution of $X_0, X_1, \ldots, X_n$.

For example, let $n = 2$ and let x_0, x_1, and x_2 each equal 0 or 1. Then

$P(X_0 = x_0, X_1 = x_1, \text{and } X_2 = x_2)$
$= P(X_0 = x_0 \text{ and } X_1 = x_1)P(X_2 = x_2 \mid X_0 = x_0 \text{ and } X_1 = x_1)$
$= P(X_0 = x_0)P(X_1 = x_1 \mid X_0 = x_0)P(X_2 = x_2 \mid X_0 = x_0 \text{ and } X_1 = x_1).$

Now $P(X_0 = x_0)$ and $P(X_1 = x_1 \mid X_0 = x_0)$ are determined by p, q, and $\pi_0(0)$; but without the Markov property, we cannot evaluate $P(X_2 = x_2 \mid X_0 = x_0 \text{ and } X_1 = x_1)$ in terms of p, q, and $\pi_0(0)$. If the Markov property is satisfied, however, then

$$P(X_2 = x_2 \mid X_0 = x_0 \text{ and } X_1 = x_1) = P(X_2 = x_2 \mid X_1 = x_1),$$

which is determined by p and q. In this case

$P(X_0 = x_0, X_1 = x_1, \text{and } X_2 = x_2)$
$= P(X_0 = x_0)P(X_1 = x_1 \mid X_0 = x_0)P(X_2 = x_2 \mid X_1 = x_1).$

For example,

$P(X_0 = 0, X_1 = 1, \text{and } X_2 = 0)$
$= P(X_0 = 0)P(X_1 = 1 \mid X_0 = 0)P(X_2 = 0 \mid X_1 = 1)$
$= \pi_0(0)pq.$

The reader should check the remaining entries in the following table, which gives the joint distribution of X_0, X_1, and X_2.

x_0	x_1	x_2	$P(X_0 = x_0, X_1 = x_1, \text{and } X_2 = x_2)$
0	0	0	$\pi_0(0)(1-p)^2$
0	0	1	$\pi_0(0)(1-p)p$
0	1	0	$\pi_0(0)pq$
0	1	1	$\pi_0(0)p(1-q)$
1	0	0	$(1-\pi_0(0))q(1-p)$
1	0	1	$(1-\pi_0(0))qp$
1	1	0	$(1-\pi_0(0))(1-q)q$
1	1	1	$(1-\pi_0(0))(1-q)^2$

1.2. Transition function and initial distribution

Let $X_n, n \geq 0$, be a Markov chain having state space $\mathscr{S}$. (The restriction to two states is now dropped.) The function $P(x, y)$, $x \in \mathscr{S}$ and $y \in \mathscr{S}$, defined by

(6) $\qquad P(x, y) = P(X_1 = y \mid X_0 = x), \qquad x, y \in \mathscr{S},$

is called the *transition function* of the chain. It is such that

(7) $\qquad P(x, y) \geq 0, \qquad x, y \in \mathscr{S},$

and

(8) $\qquad \sum_y P(x, y) = 1, \qquad x \in \mathscr{S}.$

Since the Markov chain has stationary probabilities, we see that

(9) $\qquad P(X_{n+1} = y \mid X_n = x) = P(x, y), \qquad n \geq 1.$

It now follows from the Markov property that

(10) $\quad P(X_{n+1} = y \mid X_0 = x_0, \ldots, X_{n-1} = x_{n-1}, X_n = x) = P(x, y).$

In other words, if the Markov chain is in state x at time n, then no matter how it got to x, it has probability $P(x, y)$ of being in state y at the next step. For this reason the numbers $P(x, y)$ are called the *one-step transition probabilities* of the Markov chain.

The function $\pi_0(x)$, $x \in \mathscr{S}$, defined by

(11) $\qquad \pi_0(x) = P(X_0 = x), \qquad x \in \mathscr{S},$

is called the *initial distribution* of the chain. It is such that

(12) $\qquad \pi_0(x) \geq 0, \qquad x \in \mathscr{S},$

and

(13) $\qquad \sum_x \pi_0(x) = 1.$

The joint distribution of $X_0, \ldots, X_n$ can easily be expressed in terms of the transition function and the initial distribution. For example,

$$P(X_0 = x_0, X_1 = x_1) = P(X_0 = x_0)P(X_1 = x_1 \mid X_0 = x_0)$$
$$= \pi_0(x_0)P(x_0, x_1).$$

Also,

$$P(X_0 = x_0, X_1 = x_1, X_2 = x_2)$$
$$= P(X_0 = x_0, X_1 = x_1)P(X_2 = x_2 \mid X_0 = x_0, X_1 = x_1)$$
$$= \pi_0(x_0)P(x_0, x_1)P(X_2 = x_2 \mid X_0 = x_0, X_1 = x_1).$$

Since X_n, $n \geq 0$, satisfies the Markov property and has stationary transition probabilities, we see that

$$P(X_2 = x_2 \mid X_0 = x_0, X_1 = x_1) = P(X_2 = x_2 \mid X_1 = x_1)$$
$$= P(X_1 = x_2 \mid X_0 = x_1)$$
$$= P(x_1, x_2).$$

Thus
$$P(X_0 = x_0, X_1 = x_1, X_2 = x_2) = \pi_0(x_0)P(x_0, x_1)P(x_1, x_2).$$

By induction it is easily seen that

(14) $\quad P(X_0 = x_0, \ldots, X_n = x_n) = \pi_0(x_0)P(x_0, x_1) \cdots P(x_{n-1}, x_n).$

It is usually more convenient, however, to reverse the order of our definitions. We say that $P(x, y)$, $x \in \mathscr{S}$ and $y \in \mathscr{S}$, is a *transition function* if it satisfies (7) and (8), and we say that $\pi_0(x)$, $x \in \mathscr{S}$, is an *initial distribution* if it satisfies (12) and (13). It can be shown that given any transition function P and any initial distribution π_0, there is a probability space and random variables X_n, $n \geq 0$, defined on that space satisfying (14). It is not difficult to show that these random variables form a Markov chain having transition function P and initial distribution π_0.

The reader may be bothered by the possibility that some of the conditional probabilities we have discussed may not be well defined. For example, the left side of (1) is not well defined if

$$P(X_0 = x_0, \ldots, X_n = x_n) = 0.$$

This difficulty is easily resolved. Equations (7), (8), (12), and (13) defining the transition functions and the initial distributions are well defined, and Equation (14) describing the joint distribution of $X_0, \ldots, X_n$ is well defined. It is not hard to show that if (14) holds, then (1), (6), (9), and (10) hold whenever the conditional probabilities in the respective equations are well defined. The same qualification holds for other equations involving conditional probabilities that will be obtained later.

It will soon be apparent that the transition function of a Markov chain plays a much greater role in describing its properties than does the initial distribution. For this reason it is customary to study simultaneously all Markov chains having a given transition function. In fact we adhere to the usual convention that by "a Markov chain having transition function P," we really mean the family of all Markov chains having that transition function.

1.3. Examples

In this section we will briefly describe several interesting examples of Markov chains. These examples will be further developed in the sequel.

1.3. Examples

Example 1. Random walk. Let $\xi_1, \xi_2, \ldots$ be independent integer-valued random variables having common density f. Let X_0 be an integer-valued random variable that is independent of the ξ_i's and set $X_n = X_0 + \xi_1 + \cdots + \xi_n$. The sequence X_n, $n \geq 0$, is called a *random walk*. It is a Markov chain whose state space is the integers and whose transition function is given by

$$P(x, y) = f(y - x).$$

To verify this, let π_0 denote the distribution of X_0. Then

$$\begin{aligned}
P(X_0 &= x_0, \ldots, X_n = x_n) \\
&= P(X_0 = x_0, \xi_1 = x_1 - x_0, \ldots, \xi_n = x_n - x_{n-1}) \\
&= P(X_0 = x_0) P(\xi_1 = x_1 - x_0) \cdots P(\xi_n = x_n - x_{n-1}) \\
&= \pi_0(x_0) f(x_1 - x_0) \cdots f(x_n - x_{n-1}) \\
&= \pi_0(x_0) P(x_0, x_1) \cdots P(x_{n-1}, x_n),
\end{aligned}$$

and thus (14) holds.

Suppose a "particle" moves along the integers according to this Markov chain. Whenever the particle is in x, regardless of how it got there, it jumps to state y with probability $f(y - x)$.

As a special case, consider a *simple random walk* in which $f(1) = p$, $f(-1) = q$, and $f(0) = r$, where p, q, and r are nonnegative and sum to one. The transition function is given by

$$P(x, y) = \begin{cases} p, & y = x + 1, \\ q, & y = x - 1, \\ r, & y = x, \\ 0, & \text{elsewhere.} \end{cases}$$

Let a particle undergo such a random walk. If the particle is in state x at a given observation, then by the next observation it will have jumped to state $x + 1$ with probability p and to state $x - 1$ with probability q; with probability r it will still be in state x.

Example 2. Ehrenfest chain. The following is a simple model of the exchange of heat or of gas molecules between two isolated bodies. Suppose we have two boxes, labeled 1 and 2, and d balls labeled $1, 2, \ldots, d$. Initially some of these balls are in box 1 and the remainder are in box 2. An integer is selected at random from $1, 2, \ldots, d$, and the ball labeled by that integer is removed from its box and placed in the opposite box. This procedure is repeated indefinitely with the selections being independent from trial to trial. Let X_n denote the number of balls in box 1 after the nth trial. Then X_n, $n \geq 0$, is a Markov chain on $\mathscr{S} = \{0, 1, 2, \ldots, d\}$.

The transition function of this Markov chain is easily computed. Suppose that there are x balls in box 1 at time n. Then with probability x/d the ball drawn on the $(n + 1)$th trial will be from box 1 and will be transferred to box 2. In this case there will be $x - 1$ balls in box 1 at time $n + 1$. Similarly, with probability $(d - x)/d$ the ball drawn on the $(n + 1)$th trial will be from box 2 and will be transferred to box 1, resulting in $x + 1$ balls in box 1 at time $n + 1$. Thus the transition function of this Markov chain is given by

$$P(x, y) = \begin{cases} \dfrac{x}{d}, & y = x - 1, \\ 1 - \dfrac{x}{d}, & y = x + 1, \\ 0, & \text{elsewhere.} \end{cases}$$

Note that the Ehrenfest chain can in one transition only go from state x to $x - 1$ or $x + 1$ with positive probability.

A state a of a Markov chain is called an *absorbing state* if $P(a, a) = 1$ or, equivalently, if $P(a, y) = 0$ for $y \neq a$. The next example uses this definition.

Example 3. Gambler's ruin chain. Suppose a gambler starts out with a certain initial capital in dollars and makes a series of one dollar bets against the house. Assume that he has respective probabilities p and $q = 1 - p$ of winning and losing each bet, and that if his capital ever reaches zero, he is ruined and his capital remains zero thereafter. Let X_n, $n \geq 0$, denote the gambler's capital at time n. This is a Markov chain in which 0 is an absorbing state, and for $x \geq 1$

(15) $$P(x, y) = \begin{cases} q, & y = x - 1, \\ p, & y = x + 1, \\ 0, & \text{elsewhere.} \end{cases}$$

Such a chain is called a *gambler's ruin chain* on $\mathscr{S} = \{0, 1, 2, \ldots\}$. We can modify this model by supposing that if the capital of the gambler increases to d dollars he quits playing. In this case 0 and d are both absorbing states, and (15) holds for $x = 1, \ldots, d - 1$.

For an alternative interpretation of the latter chain, we can assume that two gamblers are making a series of one dollar bets against each other and that between them they have a total capital of d dollars. Suppose the first gambler has probability p of winning any given bet, and the second gambler has probability $q = 1 - p$ of winning. The two gamblers play until one

1.3. Examples

of them goes broke. Let X_n denote the capital of the first gambler at time n. Then X_n, $n \geq 0$, is a gambler's ruin chain on $\{0, 1, \ldots, d\}$.

Example 4. Birth and death chain. Consider a Markov chain either on $\mathscr{S} = \{0, 1, 2, \ldots\}$ or on $\mathscr{S} = \{0, 1, \ldots, d\}$ such that starting from x the chain will be at $x - 1$, x, or $x + 1$ after one step. The transition function of such a chain is given by

$$P(x, y) = \begin{cases} q_x, & y = x - 1, \\ r_x, & y = x, \\ p_x, & y = x + 1, \\ 0, & \text{elsewhere,} \end{cases}$$

where p_x, q_x, and r_x are nonnegative numbers such that $p_x + q_x + r_x = 1$. The Ehrenfest chain and the two versions of the gambler's ruin chain are examples of *birth and death chains*. The phrase "birth and death" stems from applications in which the state of the chain is the population of some living system. In these applications a transition from state x to state $x + 1$ corresponds to a "birth," while a transition from state x to state $x - 1$ corresponds to a "death."

In Chapter 3 we will study *birth and death processes*. These processes are similar to birth and death chains, except that jumps are allowed to occur at arbitrary times instead of just at integer times. In most applications, the models discussed in Chapter 3 are more realistic than those obtainable by using birth and death chains.

Example 5. Queuing chain. Consider a service facility such as a checkout counter at a supermarket. People arrive at the facility at various times and are eventually served. Those customers that have arrived at the facility but have not yet been served form a waiting line or queue. There are a variety of models to describe such systems. We will consider here only one very simple and somewhat artificial model; others will be discussed in Chapter 3.

Let time be measured in convenient periods, say in minutes. Suppose that if there are any customers waiting for service at the beginning of any given period, exactly one customer will be served during that period, and that if there are no customers waiting for service at the beginning of a period, none will be served during that period. Let ξ_n denote the number of new customers arriving during the nth period. We assume that $\xi_1, \xi_2, \ldots$ are independent nonnegative integer-valued random variables having common density f.

Let X_0 denote the number of customers present initially, and for $n \geq 1$, let X_n denote the number of customers present at the end of the nth period. If $X_n = 0$, then $X_{n+1} = \xi_{n+1}$; and if $X_n \geq 1$, then $X_{n+1} = X_n + \xi_{n+1} - 1$. It follows without difficulty from the assumptions on ξ_n, $n \geq 1$, that X_n, $n \geq 0$, is a Markov chain whose state space is the nonnegative integers and whose transition function P is given by

$$P(0, y) = f(y)$$

and

$$P(x, y) = f(y - x + 1), \quad x \geq 1.$$

Example 6. Branching chain. Consider particles such as neutrons or bacteria that can generate new particles of the same type. The initial set of objects is referred to as belonging to the 0th generation. Particles generated from the nth generation are said to belong to the $(n + 1)$th generation. Let X_n, $n \geq 0$, denote the number of particles in the nth generation.

Nothing in this description requires that the various particles in a generation give rise to new particles simultaneously. Indeed at a given time, particles from several generations may coexist.

A typical situation is illustrated in Figure 1: one initial particle gives rise to two particles. Thus $X_0 = 1$ and $X_1 = 2$. One of the particles in the first generation gives rise to three particles and the other gives rise to one particle, so that $X_2 = 4$. We see from Figure 1 that $X_3 = 2$. Since neither of the particles in the third generation gives rise to new particles, we conclude that $X_4 = 0$ and consequently that $X_n = 0$ for all $n \geq 4$. In other words, the progeny of the initial particle in the zeroth generation become extinct after three generations.

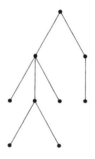

Figure 1

1.3. Examples

In order to model this system as a Markov chain, we suppose that each particle gives rise to ξ particles in the next generation, where ξ is a nonnegative integer-valued random variable having density f. We suppose that the number of offspring of the various particles in the various generations are chosen independently according to the density f.

Under these assumptions X_n, $n \geq 0$, forms a Markov chain whose state space is the nonnegative integers. State 0 is an absorbing state. For if there are no particles in a given generation, there will not be any particles in the next generation either. For $x \geq 1$

$$P(x, y) = P(\xi_1 + \cdots + \xi_x = y),$$

where $\xi_1, \ldots, \xi_x$ are independent random variables having common density f. In particular, $P(1, y) = f(y)$, $y \geq 0$.

If a particle gives rise to $\xi = 0$ particles, the interpretation is that the particle dies or disappears. Suppose a particle gives rise to ξ particles, which in turn give rise to other particles; but after some number of generations, all descendants of the initial particle have died or disappeared (see Figure 1). We describe such an event by saying that the descendants of the original particle eventually become *extinct*. An interesting problem involving branching chains is to compute the probability ρ of eventual extinction for a branching chain starting with a single particle or, equivalently, the probability that a branching chain starting at state 1 will eventually be absorbed at state 0. Once we determine ρ, we can easily find the probability that in a branching chain starting with x particles the descendants of each of the original particles eventually become extinct. Indeed, since the particles are assumed to act independently in giving rise to new particles, the desired probability is just ρ^x.

The branching chain was used originally to determine the probability that the male line of a given person would eventually become extinct. For this purpose only male children would be included in the various generations.

Example 7. Consider a gene composed of d subunits, where d is some positive integer and each subunit is either normal or mutant in form. Consider a cell with a gene composed of m mutant subunits and $d - m$ normal subunits. Before the cell divides into two daughter cells, the gene duplicates. The corresponding gene of one of the daughter cells is composed of d units chosen at random from the $2m$ mutant subunits and the $2(d - m)$ normal subunits. Suppose we follow a fixed line of descent from a given gene. Let X_0 be the number of mutant subunits initially present, and let X_n, $n \geq 1$, be the number present in the nth descendant gene. Then X_n, $n \geq 0$, is a Markov chain on $\mathscr{S} = \{0, 1, 2, \ldots, d\}$ and

$$P(x, y) = \frac{\binom{2x}{y}\binom{2d-2x}{d-y}}{\binom{2d}{d}}.$$

States 0 and d are absorbing states for this chain.

1.4. Computations with transition functions

Let X_n, $n \geq 0$, be a Markov chain on $\mathscr{S}$ having transition function P. In this section we will show how various conditional probabilities can be expressed in terms of P. We will also define the n-step transition function of the Markov chain.

We begin with the formula

(16) $\quad P(X_{n+1} = x_{n+1}, \ldots, X_{n+m} = x_{n+m} \mid X_0 = x_0, \ldots, X_n = x_n)$
$$= P(x_n, x_{n+1}) \cdots P(x_{n+m-1}, x_{n+m}).$$

To prove (16) we write the left side of this equation as

$$\frac{P(X_0 = x_0, \ldots, X_{n+m} = x_{n+m})}{P(X_0 = x_0, \ldots, X_n = x_n)}.$$

By (14) this ratio equals

$$\frac{\pi_0(x_0) P(x_0, x_1) \cdots P(x_{n+m-1}, x_{n+m})}{\pi_0(x_0) P(x_0, x_1) \cdots P(x_{n-1}, x_n)},$$

which reduces to the right side of (16).

It is convenient to rewrite (16) as

(17) $\quad P(X_{n+1} = y_1, \ldots, X_{n+m} = y_m \mid X_0 = x_0, \ldots, X_{n-1} = x_{n-1}, X_n = x)$
$$= P(x, y_1) P(y_1, y_2) \cdots P(y_{m-1}, y_m).$$

Let $A_0, \ldots, A_{n-1}$ be subsets of $\mathscr{S}$. It follows from (17) and Exercise 4(a) that

(18) $\quad P(X_{n+1} = y_1, \ldots, X_{n+m} = y_m \mid X_0 \in A_0, \ldots, X_{n-1} \in A_{n-1}, X_n = x)$
$$= P(x, y_1) P(y_1, y_2) \cdots P(y_{m-1}, y_m).$$

Let $B_1, \ldots, B_m$ be subsets of $\mathscr{S}$. It follows from (18) and Exercise 4(b) that

(19) $\quad P(X_{n+1} \in B_1, \ldots, X_{n+m} \in B_m \mid X_0 \in A_0, \ldots, X_{n-1} \in A_{n-1}, X_n = x)$
$$= \sum_{y_1 \in B_1} \cdots \sum_{y_m \in B_m} P(x, y_1) P(y_1, y_2) \cdots P(y_{m-1}, y_m).$$

1.4. Computations with transition functions

The *m-step transition function* $P^m(x, y)$, which gives the probability of going from x to y in m steps, is defined by

$$(20) \quad P^m(x, y) = \sum_{y_1} \cdots \sum_{y_{m-1}} P(x, y_1) P(y_1, y_2) \cdots P(y_{m-2}, y_{m-1}) P(y_{m-1}, y)$$

for $m \geq 2$, by $P^1(x, y) = P(x, y)$, and by

$$P^0(x, y) = \begin{cases} 1, & x = y, \\ 0, & \text{elsewhere.} \end{cases}$$

We see by setting $B_1 = \cdots = B_{m-1} = \mathscr{S}$ and $B_m = \{y\}$ in (19) that

$$(21) \quad P(X_{n+m} = y \mid X_0 \in A_0, \ldots, X_{n-1} \in A_{n-1}, X_n = x) = P^m(x, y).$$

In particular, by setting $A_0 = \cdots = A_{n-1} = \mathscr{S}$, we see that

$$(22) \quad P(X_{n+m} = y \mid X_n = x) = P^m(x, y).$$

It also follows from (21) that

$$(23) \quad P(X_{n+m} = y \mid X_0 = x, X_n = z) = P^m(z, y).$$

Since (see Exercise 4(c))

$$P^{n+m}(x, y) = P(X_{n+m} = y \mid X_0 = x)$$
$$= \sum_z P(X_n = z \mid X_0 = x) P(X_{n+m} = y \mid X_0 = x, X_n = z)$$
$$= \sum_z P^n(x, z) P(X_{n+m} = y \mid X_0 = x, X_n = z),$$

we conclude from (23) that

$$(24) \quad P^{n+m}(x, y) = \sum_z P^n(x, z) P^m(z, y).$$

For Markov chains having a finite number of states, (24) allows us to think of P^n as the nth power of the matrix P, an idea we will pursue in Section 1.4.2.

Let π_0 be an initial distribution for the Markov chain. Since

$$P(X_n = y) = \sum_x P(X_0 = x, X_n = y)$$
$$= \sum_x P(X_0 = x) P(X_n = y \mid X_0 = x),$$

we see that

$$(25) \quad P(X_n = y) = \sum_x \pi_0(x) P^n(x, y).$$

This formula allows us to compute the distribution of X_n in terms of the initial distribution π_0 and the n-step transition function P^n.

For an alternative method of computing the distribution of X_n, observe that

$$P(X_{n+1} = y) = \sum_x P(X_n = x, X_{n+1} = y)$$
$$= \sum_x P(X_n = x) P(X_{n+1} = y \mid X_n = x),$$

so that

(26) $$P(X_{n+1} = y) = \sum_x P(X_n = x) P(x, y).$$

If we know the distribution of X_0, we can use (26) to find the distribution of X_1. Then, knowing the distribution of X_1, we can use (26) to find the distribution of X_2. Similarly, we can find the distribution of X_n by applying (26) n times.

We will use the notation $P_x(\ )$ to denote probabilities of various events defined in terms of a Markov chain starting at x. Thus

$$P_x(X_1 \neq a, X_2 \neq a, X_3 = a)$$

denotes the probability that a Markov chain starting at x is in a state a at time 3 but not at time 1 or at time 2. In terms of this notation, (19) can be rewritten as

(27) $$P(X_{n+1} \in B_1, \ldots, X_{n+m} \in B_m \mid X_0 \in A_0, \ldots, X_{n-1} \in A_{n-1}, X_n = x)$$
$$= P_x(X_1 \in B_1, \ldots, X_m \in B_m).$$

1.4.1. Hitting times. Let A be a subset of $\mathscr{S}$. The *hitting time* T_A of A is defined by

$$T_A = \min(n > 0 : X_n \in A)$$

if $X_n \in A$ for some $n > 0$, and by $T_A = \infty$ if $X_n \notin A$ for all $n > 0$. In other words, T_A is the first positive time the Markov chain is in (hits) A. Hitting times play an important role in the theory of Markov chains. In this book we will be interested mainly in hitting times of sets consisting of a single point. We denote the hitting time of a point $a \in \mathscr{S}$ by T_a rather than by the more cumbersome notation $T_{\{a\}}$.

An important equation involving hitting times is given by

(28) $$P^n(x, y) = \sum_{m=1}^{n} P_x(T_y = m) P^{n-m}(y, y), \qquad n \geq 1.$$

In order to verify (28) we note that the events $\{T_y = m, X_n = y\}$, $1 \leq m \leq n$, are disjoint and that

$$\{X_n = y\} = \bigcup_{m=1}^{n} \{T_y = m, X_n = y\}.$$

1.4. Computations with transition functions

We have in effect decomposed the event $\{X_n = y\}$ according to the hitting time of y. We see from this decomposition that

$$P^n(x, y) = P_x(X_n = y)$$

$$= \sum_{m=1}^{n} P_x(T_y = m, X_n = y)$$

$$= \sum_{m=1}^{n} P_x(T_y = m)P(X_n = y \mid X_0 = x, T_y = m)$$

$$= \sum_{m=1}^{n} P_x(T_y = m)P(X_n = y \mid X_0 = x, X_1 \neq y, \ldots,$$

$$X_{m-1} \neq y, X_m = y)$$

$$= \sum_{m=1}^{n} P_x(T_y = m)P^{n-m}(y, y),$$

and hence that (28) holds.

Example 8. Show that if a is an absorbing state, then $P^n(x, a) = P_x(T_a \leq n)$, $n \geq 1$.

If a is an absorbing state, then $P^{n-m}(a, a) = 1$ for $1 \leq m \leq n$, and hence (28) implies that

$$P^n(x, a) = \sum_{m=1}^{n} P_x(T_a = m)P^{n-m}(a, a)$$

$$= \sum_{m=1}^{n} P_x(T_a = m) = P_x(T_a \leq n).$$

Observe that

$$P_x(T_y = 1) = P_x(X_1 = y) = P(x, y)$$

and that

$$P_x(T_y = 2) = \sum_{z \neq y} P_x(X_1 = z, X_2 = y) = \sum_{z \neq y} P(x, z)P(z, y).$$

For higher values of n the probabilities $P_x(T_y = n)$ can be found by using the formula

(29) $$P_x(T_y = n + 1) = \sum_{z \neq y} P(x, z)P_z(T_y = n), \quad n \geq 1.$$

This formula is a consequence of (27), but it should also be directly obvious. For in order to go from x to y for the first time at time $n + 1$, it is necessary to go to some state $z \neq y$ at the first step and then go from z to y for the first time at the end of n additional steps.

1.4.2. Transition matrix.
Suppose now that the state space $\mathscr{S}$ is finite, say $\mathscr{S} = \{0, 1, \ldots, d\}$. In this case we can think of P as the *transition matrix* having $d + 1$ rows and columns given by

$$\begin{array}{c} \quad\quad 0 \quad\quad \cdots \quad\quad d \\ \begin{matrix} 0 \\ \vdots \\ d \end{matrix} \begin{bmatrix} P(0, 0) & \cdots & P(0, d) \\ \vdots & & \vdots \\ P(d, 0) & \cdots & P(d, d) \end{bmatrix}. \end{array}$$

For example, the transition matrix of the gambler's ruin chain on $\{0, 1, 2, 3\}$ is

$$\begin{array}{c} \quad\quad 0 \quad 1 \quad 2 \quad 3 \\ \begin{matrix} 0 \\ 1 \\ 2 \\ 3 \end{matrix} \begin{bmatrix} 1 & 0 & 0 & 0 \\ q & 0 & p & 0 \\ 0 & q & 0 & p \\ 0 & 0 & 0 & 1 \end{bmatrix}. \end{array}$$

Similarly, we can regard P^n as an *n-step transition matrix*. Formula (24) with $m = n = 1$ becomes

$$P^2(x, y) = \sum_z P(x, z)P(z, y).$$

Recalling the definition of ordinary matrix multiplication, we observe that the two-step transition matrix P^2 is the product of the matrix P with itself. More generally, by setting $m = 1$ in (24) we see that

(30) $$P^{n+1}(x, y) = \sum_z P^n(x, z)P(z, y).$$

It follows from (30) by induction that the n-step transition matrix P^n is the nth power of P.

An initial distribution π_0 can be thought of as a $(d + 1)$-dimensional row vector

$$\pi_0 = (\pi_0(0), \ldots, \pi_0(d)).$$

If we let π_n denote the $(d + 1)$-dimensional row vector

$$\pi_n = (P(X_n = 0), \ldots, P(X_n = d)),$$

then (25) and (26) can be written respectively as

$$\pi_n = \pi_0 P^n$$

and

$$\pi_{n+1} = \pi_n P.$$

The two-state Markov chain discussed in Section 1.1 is one of the few examples where P^n can be found very easily.

1.5. Transient and recurrent states

Example 9. Consider the two-state Markov chain having one-step transition matrix

$$P = \begin{bmatrix} 1-p & p \\ q & 1-q \end{bmatrix},$$

where $p + q > 0$. Find P^n.

In order to find $P^n(0, 0) = P_0(X_n = 0)$, we set $\pi_0(0) = 1$ in (3) and obtain

$$P^n(0, 0) = \frac{q}{p+q} + (1-p-q)^n \frac{p}{p+q}.$$

In order to find $P^n(0, 1) = P_0(X_n = 1)$, we set $\pi_0(1) = 0$ in (4) and obtain

$$P^n(0, 1) = \frac{p}{p+q} - (1-p-q)^n \frac{p}{p+q}.$$

Similarly, we conclude that

$$P^n(1, 0) = \frac{q}{p+q} - (1-p-q)^n \frac{q}{p+q}$$

and

$$P^n(1, 1) = \frac{p}{p+q} + (1-p-q)^n \frac{q}{p+q}.$$

It follows that

$$P^n = \frac{1}{p+q} \begin{bmatrix} q & p \\ q & p \end{bmatrix} + \frac{(1-p-q)^n}{p+q} \begin{bmatrix} p & -p \\ -q & q \end{bmatrix}.$$

1.5. Transient and recurrent states

Let X_n, $n \geq 0$, be a Markov chain having state space $\mathscr{S}$ and transition function P. Set

$$\rho_{xy} = P_x(T_y < \infty).$$

Then ρ_{xy} denotes the probability that a Markov chain starting at x will be in state y at some positive time. In particular, ρ_{yy} denotes the probability that a Markov chain starting at y will ever return to y. A state y is called *recurrent* if $\rho_{yy} = 1$ and *transient* if $\rho_{yy} < 1$. If y is a recurrent state, a Markov chain starting at y returns to y with probability one. If y is a transient state, a Markov chain starting at y has positive probability $1 - \rho_{yy}$ of never returning to y. If y is an absorbing state, then $P_y(T_y = 1) =$

$P(y, y) = 1$ and hence $\rho_{yy} = 1$; thus an absorbing state is necessarily recurrent.

Let $1_y(z)$, $z \in \mathcal{S}$, denote the indicator function of the set $\{y\}$ defined by

$$1_y(z) = \begin{cases} 1, & z = y, \\ 0, & z \neq y. \end{cases}$$

Let $N(y)$ denote the number of times $n \geq 1$ that the chain is in state y. Since $1_y(X_n) = 1$ if the chain is in state y at time n and $1_y(X_n) = 0$ otherwise, we see that

(31) $$N(y) = \sum_{n=1}^{\infty} 1_y(X_n).$$

The event $\{N(y) \geq 1\}$ is the same as the event $\{T_y < \infty\}$. Thus

$$P_x(N(y) \geq 1) = P_x(T_y < \infty) = \rho_{xy}.$$

Let m and n be positive integers. By (27), the probability that a Markov chain starting at x first visits y at time m and next visits y n units of time later is $P_x(T_y = m)P_y(T_y = n)$. Thus

$$P_x(N(y) \geq 2) = \sum_{m=1}^{\infty} \sum_{n=1}^{\infty} P_x(T_y = m)P_y(T_y = n)$$

$$= \left(\sum_{m=1}^{\infty} P_x(T_y = m)\right)\left(\sum_{n=1}^{\infty} P_y(T_y = n)\right)$$

$$= \rho_{xy}\rho_{yy}.$$

Similarly we conclude that

(32) $$P_x(N(y) \geq m) = \rho_{xy}\rho_{yy}^{m-1}, \qquad m \geq 1.$$

Since

$$P_x(N(y) = m) = P_x(N(y) \geq m) - P_x(N(y) \geq m + 1),$$

it follows from (32) that

(33) $$P_x(N(y) = m) = \rho_{xy}\rho_{yy}^{m-1}(1 - \rho_{yy}), \qquad m \geq 1.$$

Also

$$P_x(N(y) = 0) = 1 - P_x(N(y) \geq 1),$$

so that

(34) $$P_x(N(y) = 0) = 1 - \rho_{xy}.$$

These formulas are intuitively obvious. To see why (33) should be true, for example, observe that a chain starting at x visits state y exactly m times if and only if it visits y for a first time, returns to y $m - 1$ additional times, and then never again returns to y.

1.5. Transient and recurrent states

We use the notation $E_x(\ )$ to denote expectations of random variables defined in terms of a Markov chain starting at x. For example,

$$(35) \qquad E_x(1_y(X_n)) = P_x(X_n = y) = P^n(x, y).$$

It follows from (31) and (35) that

$$E_x(N(y)) = E_x\left(\sum_{n=1}^{\infty} 1_y(X_n)\right)$$

$$= \sum_{n=1}^{\infty} E_x(1_y(X_n))$$

$$= \sum_{n=1}^{\infty} P^n(x, y).$$

Set

$$G(x, y) = E_x(N(y)) = \sum_{n=1}^{\infty} P^n(x, y).$$

Then $G(x, y)$ denotes the expected number of visits to y for a Markov chain starting at x.

Theorem 1 (i) *Let y be a transient state. Then*

$$P_x(N(y) < \infty) = 1$$

and

$$(36) \qquad G(x, y) = \frac{\rho_{xy}}{1 - \rho_{yy}}, \qquad x \in \mathscr{S},$$

which is finite for all $x \in \mathscr{S}$.

(ii) *Let y be a recurrent state. Then $P_y(N(y) = \infty) = 1$ and $G(y, y) = \infty$. Also*

$$(37) \qquad P_x(N(y) = \infty) = P_x(T_y < \infty) = \rho_{xy}, \qquad x \in \mathscr{S}.$$

If $\rho_{xy} = 0$, then $G(x, y) = 0$, while if $\rho_{xy} > 0$, then $G(x, y) = \infty$.

This theorem describes the fundamental difference between a transient state and a recurrent state. If y is a transient state, then no matter where the Markov chain starts, it makes only a finite number of visits to y and the expected number of visits to y is finite. Suppose instead that y is a recurrent state. Then if the Markov chain starts at y, it returns to y infinitely often. If the chain starts at some other state x, it may be impossible for it to ever hit y. If it is possible, however, and the chain does visit y at least once, then it does so infinitely often.

Proof. Let y be a transient state. Since $0 \leq \rho_{yy} < 1$, it follows from (32) that

$$P_x(N(y) = \infty) = \lim_{m \to \infty} P_x(N(y) \geq m) = \lim_{m \to \infty} \rho_{xy}\rho_{yy}^{m-1} = 0.$$

By (33)

$$G(x, y) = E_x(N(y))$$

$$= \sum_{m=1}^{\infty} m P_x(N(y) = m)$$

$$= \sum_{m=1}^{\infty} m \rho_{xy} \rho_{yy}^{m-1}(1 - \rho_{yy}).$$

Substituting $t = \rho_{yy}$ in the power series

$$\sum_{m=1}^{\infty} m t^{m-1} = \frac{1}{(1-t)^2},$$

we conclude that

$$G(x, y) = \frac{\rho_{xy}}{1 - \rho_{yy}} < \infty.$$

This completes the proof of (i).

Now let y be recurrent. Then $\rho_{yy} = 1$ and it follows from (32) that

$$P_x(N(y) = \infty) = \lim_{m \to \infty} P_x(N(y) \geq m)$$

$$= \lim_{m \to \infty} \rho_{xy} = \rho_{xy}.$$

In particular, $P_y(N(y) = \infty) = 1$. If a nonnegative random variable has positive probability of being infinite, its expectation is infinite. Thus

$$G(y, y) = E_y(N(y)) = \infty.$$

If $\rho_{xy} = 0$, then $P_x(T_y = m) = 0$ for all finite positive integers m, so (28) implies that $P^n(x, y) = 0$, $n \geq 1$; thus $G(x, y) = 0$ in this case. If $\rho_{xy} > 0$, then $P_x(N(y) = \infty) = \rho_{xy} > 0$ and hence

$$G(x, y) = E_x(N(y)) = \infty.$$

This completes the proof of Theorem 1. ∎

Let y be a transient state. Since

$$\sum_{n=1}^{\infty} P^n(x, y) = G(x, y) < \infty, \qquad x \in \mathcal{S},$$

we see that

(38) $$\lim_{n \to \infty} P^n(x, y) = 0, \qquad x \in \mathcal{S}.$$

A Markov chain is called a *transient chain* if all of its states are transient and a *recurrent chain* if all of its states are recurrent. It is easy to see that a Markov chain having a finite state space must have at least one recurrent state and hence cannot possibly be a transient chain. For if $\mathscr{S}$ is finite and all states are transient, then by (38)

$$0 = \sum_{y \in \mathscr{S}} \lim_{n \to \infty} P^n(x, y)$$

$$= \lim_{n \to \infty} \sum_{y \in \mathscr{S}} P^n(x, y)$$

$$= \lim_{n \to \infty} P_x(X_n \in \mathscr{S})$$

$$= \lim_{n \to \infty} 1 = 1,$$

which is a contradiction.

1.6. Decomposition of the state space

Let x and y be two not necessarily distinct states. We say that x *leads to* y if $\rho_{xy} > 0$. It is left as an exercise for the reader to show that x leads to y if and only if $P^n(x, y) > 0$ for some positive integer n. It is also left to the reader to show that if x leads to y and y leads to z, then x leads to z.

Theorem 2 *Let x be a recurrent state and suppose that x leads to y. Then y is recurrent and $\rho_{xy} = \rho_{yx} = 1$.*

Proof. We assume that $y \neq x$, for otherwise there is nothing to prove. Since

$$P_x(T_y < \infty) = \rho_{xy} > 0,$$

we see that $P_x(T_y = n) > 0$ for some positive integer n. Let n_0 be the least such positive integer, i.e., set

(39) $$n_0 = \min(n \geq 1 : P_x(T_y = n) > 0).$$

It follows easily from (39) and (28) that $P^{n_0}(x, y) > 0$ and

(40) $$P^m(x, y) = 0, \quad 1 \leq m < n_0.$$

Since $P^{n_0}(x, y) > 0$, we can find states $y_1, \ldots, y_{n_0-1}$ such that

$$P_x(X_1 = y_1, \ldots, X_{n_0-1} = y_{n_0-1}, X_{n_0} = y) = P(x, y_1) \cdots P(y_{n_0-1}, y) > 0.$$

None of the states $y_1, \ldots, y_{n_0-1}$ equals x or y; for if one of them did equal x or y, it would be possible to go from x to y with positive probability in fewer than n_0 steps, in contradiction to (40).

We will now show that $\rho_{yx} = 1$. Suppose on the contrary that $\rho_{yx} < 1$. Then a Markov chain starting at y has positive probability $1 - \rho_{yx}$ of never hitting x. More to the point, a Markov chain starting at x has the positive probability

$$P(x, y_1) \cdots P(y_{n_0-1}, y)(1 - \rho_{yx})$$

of visiting the states $y_1, \ldots, y_{n_0-1}, y$ successively in the first n_0 times and never returning to x after time n_0. But if this happens, the Markov chain never returns to x at any time $n \geq 1$, so we have contradicted the assumption that x is a recurrent state.

Since $\rho_{yx} = 1$, there is a positive integer n_1 such that $P^{n_1}(y, x) > 0$. Now

$$P^{n_1+n+n_0}(y, y) = P_y(X_{n_1+n+n_0} = y)$$
$$\geq P_y(X_{n_1} = x, X_{n_1+n} = x, X_{n_1+n+n_0} = y)$$
$$= P^{n_1}(y, x) P^n(x, x) P^{n_0}(x, y).$$

Hence

$$G(y, y) \geq \sum_{n=n_1+1+n_0}^{\infty} P^n(y, y)$$
$$= \sum_{n=1}^{\infty} P^{n_1+n+n_0}(y, y)$$
$$\geq P^{n_1}(y, x) P^{n_0}(x, y) \sum_{n=1}^{\infty} P^n(x, x)$$
$$= P^{n_1}(y, x) P^{n_0}(x, y) G(x, x) = +\infty,$$

from which it follows that y is also a recurrent state.

Since y is recurrent and y leads to x, we see from the part of the theorem that has already been verified that $\rho_{xy} = 1$. This completes the proof. ∎

A nonempty set C of states is said to be *closed* if no state inside of C leads to any state outside of C, i.e., if

(41) $\qquad \rho_{xy} = 0, \qquad x \in C$ and $y \notin C$.

Equivalently (see Exercise 16), C is closed if and only if

(42) $\qquad P^n(x, y) = 0, \qquad x \in C, y \notin C$, and $n \geq 1$.

Actually, even from the weaker condition

(43) $\qquad P(x, y) = 0, \qquad x \in C$ and $y \notin C$,

we can prove that C is closed. For if (43) holds, then for $x \in C$ and $y \notin C$

$$P^2(x, y) = \sum_{z \in \mathscr{S}} P(x, z) P(z, y)$$
$$= \sum_{z \in C} P(x, z) P(z, y) = 0,$$

1.6. Decomposition of the state space

and (42) follows by induction. If C is closed, then a Markov chain starting in C will, with probability one, stay in C for all time. If a is an absorbing state, then $\{a\}$ is closed.

A closed set C is called *irreducible* if x leads to y for all choices of x and y in C. It follows from Theorem 2 that if C is an irreducible closed set, then either every state in C is recurrent or every state in C is transient. The next result is an immediate consequence of Theorems 1 and 2.

Corollary 1 *Let C be an irreducible closed set of recurrent states. Then $\rho_{xy} = 1$, $P_x(N(y) = \infty) = 1$, and $G(x, y) = \infty$ for all choices of x and y in C.*

An *irreducible Markov chain* is a chain whose state space is irreducible, that is, a chain in which every state leads back to itself and also to every other state. Such a Markov chain is necessarily either a transient chain or a recurrent chain. Corollary 1 implies, in particular, that an irreducible recurrent Markov chain visits every state infinitely often with probability one.

We saw in Section 1.5 that if $\mathscr{S}$ is finite, it contains at least one recurrent state. The same argument shows that any finite closed set of states contains at least one recurrent state. Now let C be a finite irreducible closed set. We have seen that either every state in C is transient or every state in C is recurrent, and that C has at least one recurrent state. It follows that every state in C is recurrent. We summarize this result:

Theorem 3 *Let C be a finite irreducible closed set of states. Then every state in C is recurrent.*

Consider a Markov chain having a finite number of states. Theorem 3 implies that if the chain is irreducible it must be recurrent. If the chain is not irreducible, we can use Theorems 2 and 3 to determine which states are recurrent and which are transient.

Example 10. Consider a Markov chain having the transition matrix

$$\begin{array}{c} \\ 0 \\ 1 \\ 2 \\ 3 \\ 4 \\ 5 \end{array} \begin{array}{c} 012345 \\ \begin{bmatrix} 1 & 0 & 0 & 0 & 0 & 0 \\ \frac{1}{4} & \frac{1}{2} & \frac{1}{4} & 0 & 0 & 0 \\ 0 & \frac{1}{5} & \frac{2}{5} & \frac{1}{5} & 0 & \frac{1}{5} \\ 0 & 0 & 0 & \frac{1}{6} & \frac{1}{3} & \frac{1}{2} \\ 0 & 0 & 0 & \frac{1}{2} & 0 & \frac{1}{2} \\ 0 & 0 & 0 & \frac{1}{4} & 0 & \frac{3}{4} \end{bmatrix} \end{array}$$

Determine which states are recurrent and which states are transient.

As a first step in studying this Markov chain, we determine by inspection which states lead to which other states. This can be indicated in matrix form as

$$\begin{array}{c c} & \begin{array}{c c c c c c} 0 & 1 & 2 & 3 & 4 & 5 \end{array} \\ \begin{array}{c} 0 \\ 1 \\ 2 \\ 3 \\ 4 \\ 5 \end{array} & \left[\begin{array}{c c c c c c} + & 0 & 0 & 0 & 0 & 0 \\ + & + & + & + & + & + \\ + & + & + & + & + & + \\ 0 & 0 & 0 & + & + & + \\ 0 & 0 & 0 & + & + & + \\ 0 & 0 & 0 & + & + & + \end{array} \right] \end{array}$$

The x, y element of this matrix is $+$ or 0 according as ρ_{xy} is positive or zero, i.e., according as x does or does not lead to y. Of course, if $P(x, y) > 0$, then $\rho_{xy} > 0$. The converse is certainly not true in general. For example, $P(2, 0) = 0$; but

$$P^2(2, 0) = P(2, 1)P(1, 0) = \tfrac{1}{5} \cdot \tfrac{1}{4} = \tfrac{1}{20} > 0,$$

so that $\rho_{20} > 0$.

State 0 is an absorbing state, and hence also a recurrent state. We see clearly from the matrix of $+$'s and 0's that $\{3, 4, 5\}$ is an irreducible closed set. Theorem 3 now implies that 3, 4, and 5 are recurrent states. States 1 and 2 both lead to 0, but neither can be reached from 0. We see from Theorem 2 that 1 and 2 must both be transient states. In summary, states 1 and 2 are transient, and states 0, 3, 4, and 5 are recurrent.

Let $\mathscr{S}_T$ denote the collection of transient states in $\mathscr{S}$, and let $\mathscr{S}_R$ denote the collection of recurrent states in $\mathscr{S}$. In Example 10, $\mathscr{S}_T = \{1, 2\}$ and $\mathscr{S}_R = \{0, 3, 4, 5\}$. The set $\mathscr{S}_R$ can be decomposed into the disjoint irreducible closed sets $C_1 = \{0\}$ and $C_2 = \{3, 4, 5\}$. The next theorem shows that such a decomposition is always possible whenever $\mathscr{S}_R$ is nonempty.

Theorem 4 *Suppose that the set $\mathscr{S}_R$ of recurrent states is nonempty. Then $\mathscr{S}_R$ is the union of a finite or countably infinite number of disjoint irreducible closed sets $C_1, C_2, \ldots$.*

Proof. Choose $x \in \mathscr{S}_R$ and let C be the set of all states y in $\mathscr{S}_R$ such that x leads to y. Since x is recurrent, $\rho_{xx} = 1$ and hence $x \in C$. We will now verify that C is an irreducible closed set. Suppose that y is in C and y leads to z. Since y is recurrent, it follows from Theorem 2 that z is recurrent. Since x leads to y and y leads to z, we conclude that x leads to z. Thus z is in C. This shows that C is closed. Suppose that y and z are both in C. Since x is recurrent and x leads to y, it follows from

Theorem 2 that y leads to x. Since y leads to x and x leads to z, we conclude that y leads to z. This shows that C is irreducible.

To complete the proof of the theorem, we need only show that if C and D are two irreducible closed subsets of $\mathscr{S}_R$, they are either disjoint or identical. Suppose they are not disjoint and let x be in both C and D. Choose y in C. Now x leads to y, since x is in C and C is irreducible. Since D is closed, x is in D, and x leads to y, we conclude that y is in D. Thus every state in C is also in D. Similarly every state in D is also in C, so that C and D are identical. ∎

We can use our decomposition of the state space of a Markov chain to understand the behavior of such a system. If the Markov chain starts out in one of the irreducible closed sets C_i of recurrent states, it stays in C_i forever and, with probability one, visits every state in C_i infinitely often. If the Markov chain starts out in the set of transient states $\mathscr{S}_T$, it either stays in $\mathscr{S}_T$ forever or, at some time, enters one of the sets C_i and stays there from that time on, again visiting every state in that C_i infinitely often.

1.6.1 Absorption probabilities.

Let C be one of the irreducible closed sets of recurrent states, and let $\rho_C(x) = P_x(T_C < \infty)$ be the probability that a Markov chain starting at x eventually hits C. Since the chain remains permanently in C once it hits that set, we call $\rho_C(x)$ the probability that a chain starting at x is *absorbed* by the set C. Clearly $\rho_C(x) = 1$, $x \in C$, and $\rho_C(x) = 0$ if x is a recurrent state not in C. It is not so clear how to compute $\rho_C(x)$ for $x \in \mathscr{S}_T$, the set of transient states.

If there are only a finite number of transient states, and in particular if $\mathscr{S}$ itself is finite, it is always possible to compute $\rho_C(x)$, $x \in \mathscr{S}_T$, by solving a system of linear equations in which there are as many equations as unknowns, i.e., members of $\mathscr{S}_T$. To understand why this is the case, observe that if $x \in \mathscr{S}_T$, a chain starting at x can enter C only by entering C at time 1 or by being in $\mathscr{S}_T$ at time 1 and entering C at some future time. The former event has probability $\sum_{y \in C} P(x, y)$ and the latter event has probability $\sum_{y \in \mathscr{S}_T} P(x, y)\rho_C(y)$. Thus

$$(44) \quad \rho_C(x) = \sum_{y \in C} P(x, y) + \sum_{y \in \mathscr{S}_T} P(x, y)\rho_C(y), \qquad x \in \mathscr{S}_T.$$

Equation (44) holds whether $\mathscr{S}_T$ is finite or infinite, but it is far from clear how to solve (44) for the unknowns $\rho_C(x)$, $x \in \mathscr{S}_T$, when $\mathscr{S}_T$ is infinite. An additional difficulty is that if $\mathscr{S}_T$ is infinite, then (44) need not have a unique solution. Fortunately this difficulty does not arise if $\mathscr{S}_T$ is finite.

Theorem 5 *Suppose the set $\mathscr{S}_T$ of transient states is finite and let C be an irreducible closed set of recurrent states. Then the system of equations*

(45) $$f(x) = \sum_{y \in C} P(x, y) + \sum_{y \in \mathscr{S}_T} P(x, y) f(y), \qquad x \in \mathscr{S}_T,$$

has the unique solution

(46) $$f(x) = \rho_C(x), \qquad x \in \mathscr{S}_T.$$

Proof. If (45) holds, then

$$f(y) = \sum_{z \in C} P(y, z) + \sum_{z \in \mathscr{S}_T} P(y, z) f(z), \qquad y \in \mathscr{S}_T.$$

Substituting this into (45) we find that

$$f(x) = \sum_{y \in C} P(x, y) + \sum_{y \in \mathscr{S}_T} \sum_{z \in C} P(x, y) P(y, z)$$
$$+ \sum_{y \in \mathscr{S}_T} \sum_{z \in \mathscr{S}_T} P(x, y) P(y, z) f(z).$$

The sum of the first two terms is just $P_x(T_C \le 2)$, and the third term reduces to $\sum_{z \in \mathscr{S}_T} P^2(x, z) f(z)$, which is the same as $\sum_{y \in \mathscr{S}_T} P^2(x, y) f(y)$. Thus

$$f(x) = P_x(T_C \le 2) + \sum_{y \in \mathscr{S}_T} P^2(x, y) f(y).$$

By repeating this argument indefinitely or by using induction, we conclude that for all positive integers n

(47) $$f(x) = P_x(T_C \le n) + \sum_{y \in \mathscr{S}_T} P^n(x, y) f(y), \qquad x \in \mathscr{S}_T.$$

Since each $y \in \mathscr{S}_T$ is transient, it follows from (38) that

(48) $$\lim_{n \to \infty} P^n(x, y) = 0, \qquad x \in \mathscr{S} \text{ and } y \in \mathscr{S}_T.$$

According to the assumptions of the theorem, $\mathscr{S}_T$ is a finite set. It therefore follows from (48) that the sum in (47) approaches zero as $n \to \infty$. Consequently for $x \in \mathscr{S}_T$

$$f(x) = \lim_{n \to \infty} P_x(T_C \le n) = P_x(T_C < \infty) = \rho_C(x),$$

as desired. ∎

Example 11. Consider the Markov chain discussed in Example 10. Find

$$\rho_{10} = \rho_{\{0\}}(1) \quad \text{and} \quad \rho_{20} = \rho_{\{0\}}(2).$$

From (44) and the transition matrix in Example 10, we see that ρ_{10} and ρ_{20} are determined by the equations

$$\rho_{10} = \tfrac{1}{4} + \tfrac{1}{2}\rho_{10} + \tfrac{1}{4}\rho_{20}$$

and
$$\rho_{20} = \tfrac{1}{5}\rho_{10} + \tfrac{2}{5}\rho_{20}.$$

Solving these equations we find that $\rho_{10} = \tfrac{3}{5}$ and $\rho_{20} = \tfrac{1}{5}$.

By similar methods we conclude that $\rho_{\{3,4,5\}}(1) = \tfrac{2}{5}$ and $\rho_{\{3,4,5\}}(2) = \tfrac{4}{5}$. Alternatively, we can obtain these probabilities by subtracting $\rho_{\{0\}}(1)$ and $\rho_{\{0\}}(2)$ from 1, since if there are only a finite number of transient states,

(49) $$\sum_i \rho_{C_i}(x) = 1, \qquad x \in \mathscr{S}_T.$$

To verify (49) we note that for $x \in \mathscr{S}_T$

$$\sum_i \rho_{C_i}(x) = \sum_i P_x(T_{C_i} < \infty) = P_x(T_{\mathscr{S}_R} < \infty).$$

Since there are only a finite number of transient states and each transient state is visited only finitely many times, the probability $P_x(T_{\mathscr{S}_R} < \infty)$ that a recurrent state will eventually be hit is 1, so (49) holds.

Once a Markov chain starting at a transient state x enters an irreducible closed set C of recurrent states, it visits every state in C. Thus

(50) $$\rho_{xy} = \rho_C(x), \qquad x \in \mathscr{S}_T \text{ and } y \in C.$$

It follows from (50) that in our previous example

$$\rho_{13} = \rho_{14} = \rho_{15} = \rho_{\{3,4,5\}}(1) = \tfrac{2}{5}$$

and

$$\rho_{23} = \rho_{24} = \rho_{25} = \rho_{\{3,4,5\}}(2) = \tfrac{4}{5}.$$

1.6.2. Martingales. Consider a Markov chain having state space $\{0, \ldots, d\}$ and transition function P such that

(51) $$\sum_{y=0}^d yP(x, y) = x, \qquad x = 0, \ldots, d.$$

Now

$$E[X_{n+1} \mid X_0 = x_0, \ldots, X_{n-1} = x_{n-1}, X_n = x]$$

$$= \sum_{y=0}^d yP[X_{n+1} = y \mid X_0 = x_0, \ldots, X_{n-1} = x_{n-1}, X_n = x]$$

$$= \sum_{y=0}^d yP(x, y)$$

by the Markov property. We conclude from (51) that

(52) $$E[X_{n+1} \mid X_0 = x_0, \ldots, X_{n-1} = x_{n-1}, X_n = x] = x,$$

i.e., that the expected value of X_{n+1} given the past and present values of $X_0, \ldots, X_n$ equals the present value of X_n. A sequence of random variables having this property is called a *martingale*. Martingales, which need not be Markov chains, play a very important role in modern probability theory. They arose first in connection with gambling. If X_n denotes the capital of a gambler after time n and if all bets are "fair," that is, if they result in zero expected gain to the gambler, then X_n, $n \geq 0$, forms a martingale. Gamblers were naturally interested in finding some betting strategy, such as increasing their bets until they win, that would give them a net expected gain after making a series of fair bets. That this has been shown to be mathematically impossible does not seem to have deterred them from their quest.

It follows from (51) that

$$\sum_{y=0}^{d} yP(0, y) = 0,$$

and hence that $P(0, 1) = \cdots = P(0, d) = 0$. Thus 0 is necessarily an absorbing state. It follows similarly that d is an absorbing state. Consider now a Markov chain satisfying (51) and having no absorbing states other than 0 and d. It is left as an exercise for the reader to show that under these conditions the states $1, \ldots, d-1$ each lead to state 0, and hence each is a transient state. If the Markov chain starts at x, it will eventually enter one of the two absorbing states 0 and d and remain there permanently.

It follows from Example 8 that

$$E_x(X_n) = \sum_{y=0}^{d} yP_x(X_n = y)$$

$$= \sum_{y=0}^{d} yP^n(x, y)$$

$$= \sum_{y=1}^{d-1} yP^n(x, y) + dP^n(x, d)$$

$$= \sum_{y=1}^{d-1} yP^n(x, y) + dP_x(T_d \leq n).$$

Since states $1, 2, \ldots, d-1$ are transient, we see that $P^n(x, y) \to 0$ as $n \to \infty$ for $y = 1, 2, \ldots, d-1$. Consequently,

$$\lim_{n \to \infty} E_x(X_n) = dP_x(T_d < \infty) = d\rho_{xd}.$$

On the other hand, it follows from (51) (see Exercise 13(a)) that $EX_n = EX_{n-1} = \cdots = EX_0$ and hence that $E_x(X_n) = x$. Thus

$$\lim_{n \to \infty} E_x(X_n) = x.$$

1.7. Birth and death chains

By equating the two values of this limit, we conclude that

(53) $$\rho_{xd} = \frac{x}{d}, \quad x = 0, \ldots, d.$$

Since $\rho_{x0} + \rho_{xd} = 1$, it follows from (53) that

$$\rho_{x0} = 1 - \frac{x}{d}, \quad x = 0, \ldots, d.$$

Of course, once (53) is conjectured, it is easily proved directly from Theorem 5. We need only verify that for $x = 1, \ldots, d - 1$,

(54) $$\frac{x}{d} = P(x, d) + \sum_{y=1}^{d-1} \frac{y}{d} P(x, y).$$

Clearly (54) follows from (51).

The genetics chain introduced in Example 7 satisfies (51) as does a gambler's ruin chain on $\{0, 1, \ldots, d\}$ having transition matrix of the form

$$\begin{bmatrix} 1 & 0 & \cdot & \cdot & \cdot & \cdot & 0 \\ \frac{1}{2} & 0 & \frac{1}{2} & & & & \cdot \\ \cdot & \frac{1}{2} & 0 & \frac{1}{2} & & & \cdot \\ \cdot & & & & & & \cdot \\ \cdot & & & & \frac{1}{2} & 0 & \frac{1}{2} \\ 0 & \cdot & \cdot & \cdot & \cdot & 0 & 1 \end{bmatrix}$$

Suppose two gamblers make a series of one dollar bets until one of them goes broke, and suppose that each gambler has probability $\frac{1}{2}$ of winning any given bet. If the first gambler has an initial capital of x dollars and the second gambler has an initial capital of $d - x$ dollars, then the second gambler has probability $\rho_{xd} = x/d$ of going broke and the first gambler has probability $1 - (x/d)$ of going broke.

1.7. Birth and death chains

For an irreducible Markov chain either every state is recurrent or every state is transient, so that an irreducible Markov chain is either a recurrent chain or a transient chain. An irreducible Markov chain having only finitely many states is necessarily recurrent. It is generally difficult to decide whether an irreducible chain having infinitely many states is recurrent or transient. We are able to do so, however, for the birth and death chain.

Consider a birth and death chain on the nonnegative integers or on the finite set $\{0, \ldots, d\}$. In the former case we set $d = \infty$. The transition function is of the form

$$P(x, y) = \begin{cases} q_x, & y = x - 1, \\ r_x, & y = x, \\ p_x, & y = x + 1, \end{cases}$$

where $p_x + q_x + r_x = 1$ for $x \in \mathscr{S}$, $q_0 = 0$, and $p_d = 0$ if $d < \infty$. We assume additionally that p_x and q_x are positive for $0 < x < d$.

For a and b in $\mathscr{S}$ such that $a < b$, set

$$u(x) = P_x(T_a < T_b), \qquad a < x < b,$$

and set $u(a) = 1$ and $u(b) = 0$. If the birth and death chain starts at y, then in one step it goes to $y - 1$, y, or $y + 1$ with respective probabilities q_y, r_y, or p_y. It follows that

(55) $\quad u(y) = q_y u(y - 1) + r_y u(y) + p_y u(y + 1), \qquad a < y < b.$

Since $r_y = 1 - p_y - q_y$, we can rewrite (55) as

(56) $\quad u(y + 1) - u(y) = \dfrac{q_y}{p_y}(u(y) - u(y - 1)), \qquad a < y < b.$

Set $\gamma_0 = 1$ and

(57) $\quad \gamma_y = \dfrac{q_1 \cdots q_y}{p_1 \cdots p_y}, \qquad 0 < y < d.$

From (56) we see that

$$u(y + 1) - u(y) = \dfrac{\gamma_y}{\gamma_{y-1}}(u(y) - u(y - 1)), \qquad a < y < b,$$

from which it follows that

$$u(y + 1) - u(y) = \dfrac{\gamma_{a+1}}{\gamma_a} \cdots \dfrac{\gamma_y}{\gamma_{y-1}}(u(a + 1) - u(a))$$

$$= \dfrac{\gamma_y}{\gamma_a}(u(a + 1) - u(a)).$$

Consequently,

(58) $\quad u(y) - u(y + 1) = \dfrac{\gamma_y}{\gamma_a}(u(a) - u(a + 1)), \qquad a \leq y < b.$

Summing (58) on $y = a, \ldots, b - 1$ and recalling that $u(a) = 1$ and $u(b) = 0$, we conclude that

$$\dfrac{u(a) - u(a + 1)}{\gamma_a} = \dfrac{1}{\sum_{y=a}^{b-1} \gamma_y}.$$

1.7. Birth and death chains

Thus (58) becomes

$$u(y) - u(y+1) = \frac{\gamma_y}{\sum_{y=a}^{b-1} \gamma_y}, \qquad a \le y < b.$$

Summing this equation on $y = x, \ldots, b-1$ and again using the formula $u(b) = 0$, we obtain

$$u(x) = \frac{\sum_{y=x}^{b-1} \gamma_y}{\sum_{y=a}^{b-1} \gamma_y}, \qquad a < x < b.$$

It now follows from the definition of $u(x)$ that

(59) $$P_x(T_a < T_b) = \frac{\sum_{y=x}^{b-1} \gamma_y}{\sum_{y=a}^{b-1} \gamma_y}, \qquad a < x < b.$$

By subtracting both sides of (59) from 1, we see that

(60) $$P_x(T_b < T_a) = \frac{\sum_{y=a}^{x-1} \gamma_y}{\sum_{y=a}^{b-1} \gamma_y}, \qquad a < x < b.$$

Example 12. A gambler playing roulette makes a series of one dollar bets. He has respective probabilities 9/19 and 10/19 of winning and losing each bet. The gambler decides to quit playing as soon as his net winnings reach 25 dollars or his net losses reach 10 dollars.

(a) Find the probability that when he quits playing he will have won 25 dollars.

(b) Find his expected loss.

The problem fits into our scheme if we let X_n denote the capital of the gambler at time n with $X_0 = 10$. Then $X_n, n \ge 0$, forms a birth and death chain on $\{0, 1, \ldots, 35\}$ with birth and death rates

$$p_x = 9/19, \qquad 0 < x < 35,$$

and

$$q_x = 10/19, \qquad 0 < x < 35.$$

States 0 and 35 are absorbing states. Formula (60) is applicable with $a = 0$, $x = 10$, and $b = 35$. We conclude that

$$\gamma_y = (10/9)^y, \qquad 0 \le y \le 34,$$

and hence that

$$P_{10}(T_{35} < T_0) = \frac{\sum_{y=0}^{9} (10/9)^y}{\sum_{y=0}^{34} (10/9)^y} = \frac{(10/9)^{10} - 1}{(10/9)^{35} - 1} = .047.$$

Thus the gambler has probability .047 of winning 25 dollars. His expected loss in dollars is $10 - 35(.047)$, which equals $8.36.

In the remainder of this section we consider a birth and death chain on the nonnegative integers which is irreducible, i.e., such that $p_x > 0$ for $x \geq 0$ and $q_x > 0$ for $x \geq 1$. We will determine when such a chain is recurrent and when it is transient.

As a special case of (59),

(61) $$P_1(T_0 < T_n) = 1 - \frac{1}{\sum_{y=0}^{n-1} \gamma_y}, \quad n > 1.$$

Consider now a birth and death chain starting in state 1. Since the birth and death chain can move at most one step to the right at a time (considering the transition from state to state as movement along the real number line),

(62) $$1 \leq T_2 < T_3 < \cdots.$$

It follows from (62) that $\{T_0 < T_n\}$, $n > 1$, forms a nondecreasing sequence of events. We conclude from Theorem 1 of Chapter 1 of Volume I[1] that

(63) $$\lim_{n \to \infty} P_1(T_0 < T_n) = P_1(T_0 < T_n \text{ for some } n > 1).$$

Equation (62) implies that $T_n \geq n$ and thus $T_n \to \infty$ as $n \to \infty$; hence the event $\{T_0 < T_n \text{ for some } n > 1\}$ occurs if and only if the event $\{T_0 < \infty\}$ occurs. We can therefore rewrite (63) as

(64) $$\lim_{n \to \infty} P_1(T_0 < T_n) = P_1(T_0 < \infty).$$

It follows from (61) and (64) that

(65) $$P_1(T_0 < \infty) = 1 - \frac{1}{\sum_{y=0}^{\infty} \gamma_y}.$$

We are now in position to show that the birth and death chain is recurrent if and only if

(66) $$\sum_{y=0}^{\infty} \gamma_y = \infty.$$

If the birth and death chain is recurrent, then $P_1(T_0 < \infty) = 1$ and (66) follows from (65). To obtain the converse, we observe that $P(0, y) = 0$ for $y \geq 2$, and hence

(67) $$P_0(T_0 < \infty) = P(0, 0) + P(0, 1) P_1(T_0 < \infty).$$

---

[1] Paul G. Hoel, Sidney C. Port, and Charles J. Stone, *Introduction to Probability Theory* (Boston: Houghton Mifflin Co., 1971), p. 13.

Suppose (66) holds. Then by (65)

$$P_1(T_0 < \infty) = 1.$$

From this and (67) we conclude that

$$P_0(T_0 < \infty) = P(0, 0) + P(0, 1) = 1.$$

Thus 0 is a recurrent state, and since the chain is assumed to be irreducible, it must be a recurrent chain.

In summary, we have shown that an irreducible birth and death chain on $\{0, 1, 2, \ldots\}$ is recurrent if and only if

(68) $$\sum_{x=1}^{\infty} \frac{q_1 \cdots q_x}{p_1 \cdots p_x} = \infty.$$

Example 13. Consider the birth and death chain on $\{0, 1, 2, \ldots\}$ defined by

$$p_x = \frac{x+2}{2(x+1)} \quad \text{and} \quad q_x = \frac{x}{2(x+1)}, \quad x \geq 0.$$

Determine whether this chain is recurrent or transient.

Since

$$\frac{q_x}{p_x} = \frac{x}{x+2},$$

it follows that

$$\gamma_x = \frac{q_1 \cdots q_x}{p_1 \cdots p_x} = \frac{1 \cdot 2 \cdots x}{3 \cdot 4 \cdots (x+2)}$$

$$= \frac{2}{(x+1)(x+2)} = 2\left(\frac{1}{x+1} - \frac{1}{x+2}\right).$$

Thus

$$\sum_{x=1}^{\infty} \gamma_x = 2 \sum_{x=1}^{\infty} \left(\frac{1}{x+1} - \frac{1}{x+2}\right)$$

$$= 2(\tfrac{1}{2} - \tfrac{1}{3} + \tfrac{1}{3} - \tfrac{1}{4} + \tfrac{1}{4} - \tfrac{1}{5} + \cdots)$$

$$= 2 \cdot \tfrac{1}{2} = 1.$$

We conclude that the chain is transient.

1.8. Branching and queuing chains

In this section we will describe which branching chains are certain of extinction and which are not. We will also describe which queuing chains are transient and which are recurrent. The proofs of these results are somewhat complicated and will be given in the appendix to this chapter. These proofs can be skipped with no loss of continuity. It is interesting to note that the proofs of the results for the branching chain and the queuing chain are very similar, whereas the results themselves appear quite dissimilar.

1.8.1. Branching chain. Consider the branching chain introduced in Example 6. The extinction probability ρ of the chain is the probability that the descendants of a given particle eventually become extinct. Clearly

$$\rho = \rho_{10} = P_1(T_0 < \infty).$$

Suppose there are x particles present initially. Since the numbers of offspring of these particles in the various generations are chosen independently of each other, the probability ρ_{x0} that the descendants of each of the x particles eventually become extinct is just the xth power of the probability that the descendants of any one particle eventually become extinct. In other words,

(69) $$\rho_{x0} = \rho^x, \quad x = 1, 2, \ldots.$$

Recall from Example 6 that a particle gives rise to ξ particles in the next generation, where ξ is a random variable having density f. If $f(1) = 1$, the branching chain is degenerate in that every state is an absorbing state. Thus we suppose that $f(1) < 1$. Then state 0 is an absorbing state. It is left as an exercise for the reader to show that every state other than 0 is transient. From this it follows that, with probability one, the branching chain is either absorbed at 0 or approaches $+\infty$. We conclude from (69) that

$$P_x(\lim_{n \to \infty} X_n = \infty) = 1 - \rho^x, \quad x = 1, 2, \ldots.$$

Clearly it is worthwhile to determine ρ or at least to determine when $\rho = 1$ and when $\rho < 1$. This can be done using arguments based upon the formula

(70) $$\Phi(\rho) = \rho,$$

where Φ is the probability generating function of f, defined by

$$\Phi(t) = f(0) + \sum_{y=1}^{\infty} f(y) t^y, \quad 0 \le t \le 1.$$

1.8. Branching and queuing chains

To verify (70) we observe that (see Exercise 9(b))

$$\rho = \rho_{10} = P(1, 0) + \sum_{y=1}^{\infty} P(1, y)\rho_{y0}$$

$$= P(1, 0) + \sum_{y=1}^{\infty} P(1, y)\rho^y$$

$$= f(0) + \sum_{y=1}^{\infty} f(y)\rho^y$$

$$= \Phi(\rho).$$

Let μ denote the expected number of offspring of any given particle. Suppose $\mu \leq 1$. Then the equation $\Phi(t) = t$ has no roots in $[0, 1)$ (under our assumption that $f(1) < 1$), and hence $\rho = 1$. Thus *ultimate extinction is certain if $\mu \leq 1$ and $f(1) < 1$.*

Suppose instead that $\mu > 1$. Then the equation $\Phi(t) = t$ has a unique root ρ_0 in $[0, 1)$, and hence ρ equals either ρ_0 or 1. Actually ρ always equals ρ_0. Consequently, *if $\mu > 1$ the probability of ultimate extinction is less than one.*

The proofs of these results will be given in the appendix. The results themselves are intuitively very reasonable. If $\mu < 1$, then on the average each particle gives rise to fewer than one new particle, so we would expect the population to die out eventually. If $\mu > 1$, then on the average each particle gives rise to more than one new particle. In this case we would expect that the population has positive probability of growing rapidly, indeed geometrically fast, as time goes on. The case $\mu = 1$ is borderline; but since $\rho = 1$ when $\mu < 1$, it is plausible by "continuity" that $\rho = 1$ also when $\mu = 1$.

Example 14. Suppose that every man in a certain society has exactly three children, which independently have probability one-half of being a boy and one-half of being a girl. Suppose also that the number of males in the nth generation forms a branching chain. Find the probability that the male line of a given man eventually becomes extinct.

The density f of the number of male children of a given man is the binomial density with parameters $n = 3$ and $p = \frac{1}{2}$. Thus $f(0) = \frac{1}{8}$, $f(1) = \frac{3}{8}, f(2) = \frac{3}{8}, f(3) = \frac{1}{8}$, and $f(x) = 0$ for $x \geq 4$. The mean number of male offspring is $\mu = \frac{3}{2}$. Since $\mu > 1$, the extinction probability ρ is the root of the equation

$$\tfrac{1}{8} + \tfrac{3}{8}t + \tfrac{3}{8}t^2 + \tfrac{1}{8}t^3 = t$$

lying in $[0, 1)$. We can rewrite this equation as
$$t^3 + 3t^2 - 5t + 1 = 0,$$
or equivalently as
$$(t - 1)(t^2 + 4t - 1) = 0.$$
This equation has three roots, namely, 1, $-\sqrt{5} - 2$, and $\sqrt{5} - 2$. Consequently, $\rho = \sqrt{5} - 2$.

1.8.2. Queuing chain. Consider the queuing chain introduced in Example 5. Let $\xi_1, \xi_2, \ldots$ and μ be as in that example. In this section we will indicate when the queuing chain is recurrent and when it is transient.

Let μ denote the expected number of customers arriving in unit time. Suppose first that $\mu > 1$. Since at most one person is served at a time and on the average more than one new customer enters the queue at a time, it would appear that as time goes on more and more people will be waiting for service and that the queue length will approach infinity. This is indeed the case, so that *if $\mu > 1$ the queuing chain is transient*.

In discussing the case $\mu \leq 1$, we will assume that the chain is irreducible (see Exercises 37 and 38 for necessary and sufficient conditions for irreducibility and for results when the queuing chain is not irreducible). Suppose first that $\mu < 1$. Then on the average fewer than one new customer will enter the queue in unit time. Since one customer is served whenever the queue is nonempty, we would expect that, regardless of the initial length of the queue, it will become empty at some future time. This is indeed the case and, in particular, 0 is a recurrent state. The case $\mu = 1$ is borderline, but again it turns out that 0 is a recurrent state. Thus *if $\mu \leq 1$ and the queuing chain is irreducible, it is recurrent*.

The proof of these results will be given in the appendix.

APPENDIX

1.9. Proof of results for the branching and queuing chains

In this section we will verify the results discussed in Section 1.8. To do so we need the following.

Theorem 6 *Let Φ be the probability generating function of a nonnegative integer-valued random variable ξ and set $\mu = E\xi$ (with $\mu = +\infty$ if ξ does not have finite expectation). If $\mu \leq 1$ and $P(\xi = 1) < 1$, the equation*

(71) $$\Phi(t) = t$$

has no roots in $[0, 1)$. If $\mu > 1$, then (71) has a unique root ρ_0 in $[0, 1)$.

1.9. Proof of results for the branching and queuing chains

Graphs of $\Phi(t)$, $0 \le t \le 1$, in three typical cases corresponding to $\mu < 1$, $\mu = 1$, and $\mu > 1$ are shown in Figure 2. The fact that μ is the left-hand derivative of $\Phi(t)$ at $t = 1$ plays a fundamental role in the proof of Theorem 6.

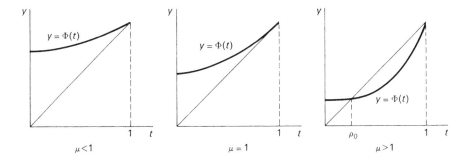

Figure 2

Proof. Let f denote the density of ξ. Then

$$\Phi(t) = f(0) + f(1)t + f(2)t^2 + \cdots$$

and

$$\Phi'(t) = f(1) + 2f(2)t + 3f(3)t^2 + \cdots.$$

Thus $\Phi(0) = f(0)$, $\Phi(1) = 1$, and

$$\lim_{t \to 1} \Phi'(t) = f(1) + 2f(2) + 3f(3) + \cdots = \mu.$$

Suppose first that $\mu < 1$. Then

$$\lim_{t \to 1} \Phi'(t) < 1.$$

Since $\Phi'(t)$ is nondecreasing in t, $0 \le t < 1$, we conclude that $\Phi'(t) < 1$ for $0 \le t < 1$. Suppose next that $\mu = 1$ and $f(1) = P(\xi = 1) < 1$. Then $f(n) > 0$ for some $n \ge 2$ (otherwise $f(0) > 0$, which implies that $\mu < 1$, a contradiction). Therefore $\Phi'(t)$ is strictly increasing in t, $0 \le t < 1$. Since

$$\lim_{t \to 1} \Phi'(t) = 1,$$

we again conclude that $\Phi'(t) < 1$ for $0 \le t < 1$.

Suppose now that $\mu \le 1$ and $P(\xi = 1) < 1$. We have shown that $\Phi'(t) < 1$ for $0 \le t < 1$. Thus

$$\frac{d}{dt}(\Phi(t) - t) < 0, \quad 0 \le t < 1,$$

and hence $\Phi(t) - t$ is strictly decreasing on $[0, 1]$. Since $\Phi(1) - 1 = 0$, we see that $\Phi(t) - t > 0, 0 \leq t < 1$, and hence that (71) has no roots on $[0, 1)$. This proves the first part of the theorem.

Suppose next that $\mu > 1$. Then
$$\lim_{t \to 1} \Phi'(t) > 1,$$
so by the continuity of Φ' there is a number t_0 such that $0 < t_0 < 1$ and $\Phi'(t) > 1$ for $t_0 < t < 1$. It follows from the mean value theorem that
$$\frac{\Phi(1) - \Phi(t_0)}{1 - t_0} > 1.$$
Since $\Phi(1) = 1$, we conclude that $\Phi(t_0) - t_0 < 0$. Now $\Phi(t) - t$ is continuous in t and nonnegative at $t = 0$, so by the intermediate value theorem it must have a zero ρ_0 on $[0, t_0)$. Thus (71) has a root ρ_0 in $[0, 1)$. We will complete the proof of the theorem by showing that there is only one such root.

Suppose that $0 \leq \rho_0 < \rho_1 < 1$, $\Phi(\rho_0) = \rho_0$, and $\Phi(\rho_1) = \rho_1$. Then the function $\Phi(t) - t$ vanishes at ρ_0, ρ_1, and 1; hence by Rolle's theorem its first derivative has at least two roots in $(0, 1)$. By another application of Rolle's theorem its second derivative $\Phi''(t)$ has at least one root in $(0, 1)$. But if $\mu > 1$, then at least one of the numbers $f(2), f(3), \ldots$ is strictly positive, and hence
$$\Phi''(t) = 2f(2) + 3 \cdot 2f(3)t + \cdots$$
has no roots in $(0, 1)$. This contradiction shows that $\Phi(t) = t$ has a unique root in $[0, 1)$. ∎

1.9.1. Branching chain. Using Theorem 6 we see that the results for $\mu \leq 1$ follow as indicated in Section 1.8.1.

Suppose $\mu > 1$. It follows from Theorem 6 that ρ equals ρ_0 or 1, where ρ_0 is the unique root of the equation $\Phi(t) = t$ in $[0, 1)$. We will show that ρ always equals ρ_0.

First we observe that since the initial particles act independently in giving rise to their offspring, the probability $P_y(T_0 \leq n)$ that the descendants of each of the $y \geq 1$ particles become extinct by time n is given by
$$P_y(T_0 \leq n) = (P_1(T_0 \leq n))^y.$$
Consequently for $n \geq 0$ by Exercise 9(a)
$$P_1(T_0 \leq n + 1) = P(1, 0) + \sum_{y=1}^{\infty} P(1, y) P_y(T_0 \leq n)$$
$$= P(1, 0) + \sum_{y=1}^{\infty} P(1, y)(P_1(T_0 \leq n))^y$$
$$= f(0) + \sum_{y=1}^{\infty} f(y)(P_1(T_0 \leq n))^y,$$

1.9. Proof of results for the branching and queuing chains

and hence

(72) $\quad P_1(T_0 \leq n+1) = \Phi(P_1(T_0 \leq n)), \quad n \geq 0.$

We will use (72) to prove by induction that

(73) $\quad P_1(T_0 \leq n) \leq \rho_0, \quad n \geq 0.$

Now

$$P_1(T_0 \leq 0) = 0 \leq \rho_0,$$

so that (73) is true for $n = 0$. Suppose that (73) holds for a given value of n. Since $\Phi(t)$ is increasing in t, we conclude from (72) that

$$P_1(T_0 \leq n+1) = \Phi(P_1(T_0 \leq n)) \leq \Phi(\rho_0) = \rho_0,$$

and thus (73) holds for the next value of n. By induction (73) is true for all $n \geq 0$.

By letting $n \to \infty$ in (73) we see that

$$\rho = P_1(T_0 < \infty) = \lim_{n \to \infty} P_1(T_0 \leq n) \leq \rho_0.$$

Since ρ is one of the two numbers ρ_0 or 1, it must be the number ρ_0.

1.9.2. Queuing chain. We will now verify the results of Section 1.8.2. Let ξ_n denote the number of customers arriving during the nth time period. Then $\xi_1, \xi_2, \ldots$ are independent random variables having common density f, mean μ, and probability generating function Φ.

It follows from Exercise 9(b) and the identity $P(0, z) \equiv P(1, z)$, valid for a queuing chain, that $\rho_{00} = \rho_{10}$. We will show that the number $\rho = \rho_{00} = \rho_{10}$ satisfies the equation

(74) $\quad \Phi(\rho) = \rho.$

If 0 is a recurrent state, $\rho = 1$ and (74) follows immediately from the fact that $\Phi(1) = 1$. To verify (74) in general, we observe first that by Exercise 9(b)

$$\rho_{00} = P(0, 0) + \sum_{y=1}^{\infty} P(0, y) \rho_{y0},$$

i.e., that

(75) $\quad \rho = f(0) + \sum_{y=1}^{\infty} f(y) \rho_{y0}.$

In order to compute ρ_{y0}, $y = 1, 2, \ldots$, we consider a queuing chain starting at the positive integer y. For $n = 1, 2, \ldots$, the event $\{T_{y-1} = n\}$ occurs if and only if

$$n = \min\,(m > 0 \colon y + (\xi_1 - 1) + \cdots + (\xi_m - 1) = y - 1)$$
$$= \min\,(m > 0 \colon \xi_1 + \cdots + \xi_m = m - 1),$$

that is, if and only if n is the smallest positive integer m such that the number of new customers entering the queue by time m is one less than the number served by time m. Thus $P_y(T_{y-1} = n)$ is independent of y, and consequently $\rho_{y,y-1} = P_y(T_{y-1} < \infty)$ is independent of y for $y = 1, 2, \ldots$. Since $\rho_{10} = \rho$, we see that

$$\rho_{y,y-1} = \rho_{y-1,y-2} = \cdots = \rho_{10} = \rho.$$

Now the queuing chain can go at most one step to the left at a time, so in order to go from state $y > 0$ to state 0 it must pass through all the intervening states $y - 1, \ldots, 1$. By applying the Markov property we can conclude (see Exercise 39) that

(76) $$\rho_{y0} = \rho_{y,y-1}\rho_{y-1,y-2}\cdots\rho_{10} = \rho^y.$$

It follows from (75) and (76) that

$$\rho = f(0) + \sum_{y=1}^{\infty} f(y)\rho^y = \Phi(\rho),$$

so that (74) holds.

Using (74) and Theorem 6 it is easy to see that if $\mu \leq 1$ and the queuing chain is irreducible, then the chain is recurrent. For ρ satisfies (74) and by Theorem 6 this equation has no roots in $[0, 1)$ (observe that $P(\xi_1 = 1) < 1$ if the queuing chain is irreducible). We conclude that $\rho = 1$. Since $\rho_{00} = \rho$, state 0 is recurrent, and thus since the chain is irreducible, all states are recurrent.

Suppose now that $\mu > 1$. Again ρ satisfies (74) which, by Theorem 6, has a unique root ρ_0 in $[0, 1)$. Thus ρ equals either ρ_0 or 1. We will prove that $\rho = \rho_0$.

To this end we first observe that by Exercise 9(a)

$$P_1(T_0 \leq n + 1) = P(1, 0) + \sum_{y=1}^{\infty} P(1, y)P_y(T_0 \leq n),$$

which can be rewritten as

(77) $$P_1(T_0 \leq n + 1) = f(0) + \sum_{y=1}^{\infty} f(y)P_y(T_0 \leq n).$$

We claim next that

(78) $$P_y(T_0 \leq n) \leq (P_1(T_0 \leq n))^y, \quad y \geq 1 \text{ and } n \geq 0.$$

To verify (78) observe that if a queuing chain starting at y reaches 0 in n or fewer steps, it must reach $y - 1$ in n or fewer steps, go from $y - 1$ to $y - 2$ in n or fewer steps, etc. By applying the Markov property we can conclude (see Exercise 39) that

(79) $$P_y(T_0 \leq n) \leq P_y(T_{y-1} \leq n)P_{y-1}(T_{y-2} \leq n) \cdots P_1(T_0 \leq n).$$

Since
$$P_z(T_{z-1} \le n) = P_1(T_0 \le n), \quad 1 \le z \le y,$$

(78) is valid.

It follows from (77) and (78) that
$$P_1(T_0 \le n+1) \le f(0) + \sum_{y=1}^{\infty} f(y)(P_1(T_0 \le n))^y,$$

i.e., that

(80) $$P_1(T_0 \le n+1) \le \Phi(P_1(T_0 \le n)), \quad n \ge 0.$$

This in turn implies that

(81) $$P_1(T_0 \le n) \le \rho_0, \quad n \ge 0,$$

by a proof that is almost identical to the proof that (72) implies (73) (the slight changes needed are left as an exercise for the reader). Just as in the proof of the corresponding result for the branching chain, we see by letting $n \to \infty$ in (81) that $\rho \le \rho_0$ and hence that $\rho = \rho_0$.

We have shown that if $\mu > 1$, then $\rho_{00} = \rho < 1$, and hence 0 is a transient state. It follows that if $\mu > 1$ and the chain is irreducible, then all states are transient. If $\mu > 1$ and the queuing chain is not irreducible, then case (d) of Exercise 38 holds (why?), and it is left to the reader to show that again all states are transient.

Exercises

1 Let X_n, $n \ge 0$, be the two-state Markov chain. Find
  (a) $P(X_1 = 0 \mid X_0 = 0 \text{ and } X_2 = 0)$,
  (b) $P(X_1 \ne X_2)$.

2 Suppose we have two boxes and $2d$ balls, of which d are black and d are red. Initially, d of the balls are placed in box 1, and the remainder of the balls are placed in box 2. At each trial a ball is chosen at random from each of the boxes, and the two balls are put back in the opposite boxes. Let X_0 denote the number of black balls initially in box 1 and, for $n \ge 1$, let X_n denote the number of black balls in box 1 after the nth trial. Find the transition function of the Markov chain X_n, $n \ge 0$.

3 Let the queuing chain be modified by supposing that if there are one or more customers waiting to be served at the start of a period, there is probability p that one customer will be served during that period and probability $1 - p$ that no customers will be served during that period. Find the transition function for this modified queuing chain.

4 Consider a probability space $(\Omega, \mathcal{A}, P)$ and assume that the various sets mentioned below are all in $\mathcal{A}$.
  (a) Show that if D_i are disjoint and $P(C \mid D_i) = p$ independently of i, then $P(C \mid \bigcup_i D_i) = p$.
  (b) Show that if C_i are disjoint, then $P(\bigcup_i C_i \mid D) = \sum_i P(C_i \mid D)$.
  (c) Show that if E_i are disjoint and $\bigcup_i E_i = \Omega$, then
$$P(C \mid D) = \sum_i P(E_i \mid D) P(C \mid E_i \cap D).$$
  (d) Show that if C_i are disjoint and $P(A \mid C_i) = P(B \mid C_i)$ for all i, then $P(A \mid \bigcup_i C_i) = P(B \mid \bigcup_i C_i)$.

5 Let X_n, $n \geq 0$, be the two-state Markov chain.
  (a) Find $P_0(T_0 = n)$.
  (b) Find $P_0(T_1 = n)$.

6 Let X_n, $n \geq 0$, be the Ehrenfest chain and suppose that X_0 has a binomial distribution with parameters d and $1/2$, i.e.,
$$P(X_0 = x) = \frac{\binom{d}{x}}{2^d}, \qquad x = 0, \ldots, d.$$
Find the distribution of X_1.

7 Let X_n, $n \geq 0$, be a Markov chain. Show that
$$P(X_0 = x_0 \mid X_1 = x_1, \ldots, X_n = x_n) = P(X_0 = x_0 \mid X_1 = x_1).$$

8 Let x and y be distinct states of a Markov chain having $d < \infty$ states and suppose that x leads to y. Let n_0 be the smallest positive integer such that $P^{n_0}(x, y) > 0$ and let $x_1, \ldots, x_{n_0 - 1}$ be states such that
$$P(x, x_1) P(x_1, x_2) \cdots P(x_{n_0 - 2}, x_{n_0 - 1}) P(x_{n_0 - 1}, y) > 0.$$
  (a) Show that $x, x_1, \ldots, x_{n_0 - 1}, y$ are distinct states.
  (b) Use (a) to show that $n_0 \leq d - 1$.
  (c) Conclude that $P_x(T_y \leq d - 1) > 0$.

9 Use (29) to verify the following identities:
  (a) $P_x(T_y \leq n + 1) = P(x, y) + \sum_{z \neq y} P(x, z) P_z(T_y \leq n)$, $n \geq 0$;
  (b) $\rho_{xy} = P(x, y) + \sum_{z \neq y} P(x, z) \rho_{zy}$.

10 Consider the Ehrenfest chain with $d = 3$.
  (a) Find $P_x(T_0 = n)$ for $x \in \mathcal{S}$ and $1 \leq n \leq 3$.
  (b) Find P, P^2, and P^3.
  (c) Let π_0 be the uniform distribution $\pi_0 = (\frac{1}{4}, \frac{1}{4}, \frac{1}{4}, \frac{1}{4})$. Find π_1, π_2, and π_3.

11 Consider the genetics chain from Example 7 with $d = 3$.
  (a) Find the transition matrices P and P^2.
  (b) If $\pi_0 = (0, \tfrac{1}{2}, \tfrac{1}{2}, 0)$, find π_1 and π_2.
  (c) Find $P_x(T_{\{0,3\}} = n)$, $x \in \mathcal{S}$, for $n = 1$ and $n = 2$.

12 Consider the Markov chain having state space $\{0, 1, 2\}$ and transition matrix

$$P = \begin{matrix} \\ 0 \\ 1 \\ 2 \end{matrix} \begin{matrix} 0 & 1 & 2 \end{matrix} \begin{bmatrix} 0 & 1 & 0 \\ 1-p & 0 & p \\ 0 & 1 & 0 \end{bmatrix}.$$

  (a) Find P^2.
  (b) Show that $P^4 = P^2$.
  (c) Find P^n, $n \geq 1$.

13 Let X_n, $n \geq 0$, be a Markov chain whose state space $\mathcal{S}$ is a subset of $\{0, 1, 2, \ldots\}$ and whose transition function P is such that

$$\sum_y y P(x, y) = Ax + B, \qquad x \in \mathcal{S},$$

for some constants A and B.
  (a) Show that $EX_{n+1} = AEX_n + B$.
  (b) Show that if $A \neq 1$, then

$$EX_n = \frac{B}{1-A} + A^n \left(EX_0 - \frac{B}{1-A} \right).$$

14 Let X_n, $n \geq 0$, be the Ehrenfest chain on $\{0, 1, \ldots, d\}$. Show that the assumption of Exercise 13 holds and use that exercise to compute $E_x(X_n)$.

15 Let y be a transient state. Use (36) to show that for all x

$$\sum_{n=0}^{\infty} P^n(x, y) \leq \sum_{n=0}^{\infty} P^n(y, y).$$

16 Show that $\rho_{xy} > 0$ if and only if $P^n(x, y) > 0$ for some positive integer n.

17 Show that if x leads to y and y leads to z, then x leads to z.

18 Consider a Markov chain on the nonnegative integers such that, starting from x, the chain goes to state $x + 1$ with probability p, $0 < p < 1$, and goes to state 0 with probability $1 - p$.
  (a) Show that this chain is irreducible.
  (b) Find $P_0(T_0 = n)$, $n \geq 1$.
  (c) Show that the chain is recurrent.

19 Consider a Markov chain having state space $\{0, 1, \ldots, 6\}$ and transition matrix

$$\begin{array}{c} \\ 0 \\ 1 \\ 2 \\ 3 \\ 4 \\ 5 \\ 6 \end{array} \begin{array}{cccccccc} 0 & 1 & 2 & 3 & 4 & 5 & 6 \\ \left[\begin{array}{ccccccc} \frac{1}{2} & 0 & \frac{1}{8} & \frac{1}{4} & \frac{1}{8} & 0 & 0 \\ 0 & 0 & 1 & 0 & 0 & 0 & 0 \\ 0 & 0 & 0 & 1 & 0 & 0 & 0 \\ 0 & 1 & 0 & 0 & 0 & 0 & 0 \\ 0 & 0 & 0 & 0 & \frac{1}{2} & 0 & \frac{1}{2} \\ 0 & 0 & 0 & 0 & \frac{1}{2} & \frac{1}{2} & 0 \\ 0 & 0 & 0 & 0 & 0 & \frac{1}{2} & \frac{1}{2} \end{array}\right]. \end{array}$$

(a) Determine which states are transient and which states are recurrent.
(b) Find ρ_{0y}, $y = 0, \ldots, 6$.

20 Consider the Markov chain on $\{0, 1, \ldots, 5\}$ having transition matrix

$$\begin{array}{c} \\ 0 \\ 1 \\ 2 \\ 3 \\ 4 \\ 5 \end{array} \begin{array}{cccccc} 0 & 1 & 2 & 3 & 4 & 5 \\ \left[\begin{array}{cccccc} \frac{1}{2} & \frac{1}{2} & 0 & 0 & 0 & 0 \\ \frac{1}{3} & \frac{2}{3} & 0 & 0 & 0 & 0 \\ 0 & 0 & \frac{1}{8} & 0 & \frac{7}{8} & 0 \\ \frac{1}{4} & \frac{1}{4} & 0 & 0 & \frac{1}{4} & \frac{1}{4} \\ 0 & 0 & \frac{3}{4} & 0 & \frac{1}{4} & 0 \\ 0 & \frac{1}{5} & 0 & \frac{1}{5} & \frac{1}{5} & \frac{2}{5} \end{array}\right]. \end{array}$$

(a) Determine which states are transient and which are recurrent.
(b) Find $\rho_{\{0,1\}}(x)$, $x = 0, \ldots, 5$.

21 Consider a Markov chain on $\{0, 1, \ldots, d\}$ satisfying (51) and having no absorbing states other than 0 and d. Show that the states $1, \ldots, d-1$ each lead to 0, and hence that each is a transient state.

22 Show that the genetics chain introduced in Example 7 satisfies Equation (51).

23 A certain Markov chain that arises in genetics has states $0, 1, \ldots, 2d$ and transition function

$$P(x, y) = \binom{2d}{y} \left(\frac{x}{2d}\right)^y \left(1 - \frac{x}{2d}\right)^{2d-y}.$$

Find $\rho_{\{0\}}(x)$, $0 < x < 2d$.

24 Consider a gambler's ruin chain on $\{0, 1, \ldots, d\}$. Find

$$P_x(T_0 < T_d), \qquad 0 < x < d.$$

25 A gambler playing roulette makes a series of one dollar bets. He has respective probabilities 9/19 and 10/19 of winning and losing each bet. The gambler decides to quit playing as soon as he either is one dollar ahead or has lost his initial capital of $1000.
(a) Find the probability that when he quits playing he will have lost $1000.
(b) Find his expected loss.

26 Consider a birth and death chain on the nonnegative integers such that $p_x > 0$ and $q_x > 0$ for $x \geq 1$.
  (a) Show that if $\sum_{y=0}^{\infty} \gamma_y = \infty$, then $\rho_{x0} = 1$, $x \geq 1$.
  (b) Show that if $\sum_{y=0}^{\infty} \gamma_y < \infty$, then

$$\rho_{x0} = \frac{\sum_{y=x}^{\infty} \gamma_y}{\sum_{y=0}^{\infty} \gamma_y}, \qquad x \geq 1.$$

27 Consider a gambler's ruin chain on $\{0, 1, 2, \ldots\}$.
  (a) Show that if $q \geq p$, then $\rho_{x0} = 1$, $x \geq 1$.
  (b) Show that if $q < p$, then $\rho_{x0} = (q/p)^x$, $x \geq 1$.
  Hint: Use Exercise 26.

28 Consider an irreducible birth and death chain on the nonnegative integers. Show that if $p_x \leq q_x$ for $x \geq 1$, the chain is recurrent.

29 Consider an irreducible birth and death chain on the nonnegative integers such that

$$\frac{q_x}{p_x} = \left(\frac{x}{x+1}\right)^2, \qquad x \geq 1.$$

  (a) Show that this chain is transient.
  (b) Find ρ_{x0}, $x \geq 1$. Hint: Use Exercise 26 and the formula $\sum_{y=1}^{\infty} 1/y^2 = \pi^2/6$.

30 Consider the birth and death chain in Example 13.
  (a) Compute $P_x(T_a < T_b)$ for $a < x < b$.
  (b) Compute ρ_{x0}, $x > 0$.

31 Consider a branching chain such that $f(1) < 1$. Show that every state other than 0 is transient.

32 Consider the branching chain described in Example 14. If a given man has two boys and one girl, what is the probability that his male line will continue forever?

33 Consider a branching chain with $f(0) = f(3) = 1/2$. Find the probability ρ of extinction.

34 Consider a branching chain with $f(x) = p(1-p)^x$, $x \geq 0$, where $0 < p < 1$. Show that $\rho = 1$ if $p \geq 1/2$ and that $\rho = p/(1-p)$ if $p < 1/2$.

35 Let X_n, $n \geq 0$, be a branching chain. Show that $E_x(X_n) = x\mu^n$. Hint: See Exercise 13.

36 Let X_n, $n \geq 0$, be a branching chain and suppose that the associated random variable ξ has finite variance σ^2.
  (a) Show that

$$E[X_{n+1}^2 \mid X_n = x] = x\sigma^2 + x^2\mu^2.$$

  (b) Use Exercise 35 to show that

$$E_x(X_{n+1}^2) = x\mu^n\sigma^2 + \mu^2 E_x(X_n^2).$$

  Hint: Use the formula $EY = \sum_x P(X = x) E[Y \mid X = x]$.

(c) Show that
$$E_x(X_n^2) = x\sigma^2(\mu^{n-1} + \cdots + \mu^{2(n-1)}) + x^2\mu^{2n}, \quad n \geq 1.$$
(d) Show that if there are x particles initially, then for $n \geq 1$
$$\text{Var } X_n = \begin{cases} x\sigma^2\mu^{n-1}\left(\dfrac{1-\mu^n}{1-\mu}\right), & \mu \neq 1, \\ nx\sigma^2, & \mu = 1. \end{cases}$$

37 Consider the queuing chain.
(a) Show that if either $f(0) = 0$ or $f(0) + f(1) = 1$, the chain is not irreducible.
(b) Show that if $f(0) > 0$ and $f(0) + f(1) < 1$, the chain is irreducible. *Hint:* First verify that (i) $p_{xy} > 0$ for $0 \leq y < x$; and (ii) if $x_0 \geq 2$ and $f(x_0) > 0$, then $p_{0, x_0 + n(x_0 - 1)} > 0$ for $n \geq 0$.

38 Determine which states of the queuing chain are absorbing, which are recurrent, and which are transient, when the chain is not irreducible. Consider the following four cases separately (see Exercise 37):
(a) $f(1) = 1$;
(b) $f(0) > 0, f(1) > 0,$ and $f(0) + f(1) = 1$;
(c) $f(0) = 1$;
(d) $f(0) = 0$ and $f(1) < 1$.

39 Consider the queuing chain.
(a) Show that for $y \geq 2$ and m a positive integer
$$P_y(T_0 = m) = \sum_{k=1}^{m-1} P_y(T_{y-1} = k) P_{y-1}(T_0 = m - k).$$
(b) By summing the equation in (a) on $m = 1, 2, \ldots$, show that
$$\rho_{y0} = \rho_{y, y-1} \rho_{y-1, 0} \quad y \geq 2.$$
(c) Why does Equation (76) follow from (b)?
(d) By summing the equation in (a) on $m = 1, 2, \ldots, n$, show that
$$P_y(T_0 \leq n) \leq P_y(T_{y-1} \leq n) P_{y-1}(T_0 \leq n), \quad y \geq 2.$$
(e) Why does Equation (79) follow from (d)?

40 Verify that (81) follows from (80) by induction.

2 Stationary Distributions of a Markov Chain

Let X_n, $n \geq 0$, be a Markov chain having state space $\mathscr{S}$ and transition function P. If $\pi(x)$, $x \in \mathscr{S}$, are nonnegative numbers summing to one, and if

(1) $$\sum_x \pi(x) P(x, y) = \pi(y), \quad y \in \mathscr{S},$$

then π is called a *stationary distribution*. Suppose that a stationary distribution π exists and that

(2) $$\lim_{n \to \infty} P^n(x, y) = \pi(y), \quad y \in \mathscr{S}.$$

Then, as we will soon see, regardless of the initial distribution of the chain, the distribution of X_n approaches π as $n \to \infty$. In such cases, π is sometimes called the *steady state distribution*.

In this chapter we will determine which Markov chains have stationary distributions, when there is such a unique distribution, and when (2) holds.

2.1. Elementary properties of stationary distributions

Let π be a stationary distribution. Then

$$\sum_x \pi(x) P^2(x, y) = \sum_x \pi(x) \sum_z P(x, z) P(z, y)$$

$$= \sum_z \left(\sum_x \pi(x) P(x, z) \right) P(z, y)$$

$$= \sum_z \pi(z) P(z, y) = \pi(y).$$

Similarly by induction based on the formula

$$P^{n+1}(x, y) = \sum_z P^n(x, z) P(z, y),$$

we conclude that for all n

(3) $$\sum_x \pi(x) P^n(x, y) = \pi(y), \quad y \in \mathscr{S}.$$

If X_0 has the stationary distribution π for its initial distribution, then (3) implies that for all n

(4) $$P(X_n = y) = \pi(y), \qquad y \in \mathscr{S},$$

and hence that the distribution of X_n is independent of n. Suppose conversely that the distribution of X_n is independent of n. Then the initial distribution π_0 is such that

$$\pi_0(y) = P(X_0 = y) = P(X_1 = y) = \sum_x \pi_0(x) P(x, y).$$

Consequently π_0 is a stationary distribution. In summary, the distribution of X_n is independent of n if and only if the initial distribution is a stationary distribution.

Suppose now that π is a stationary distribution and that (2) holds. Let π_0 be the initial distribution. Then

(5) $$P(X_n = y) = \sum_x \pi_0(x) P^n(x, y), \qquad y \in \mathscr{S}.$$

By using (2) and the bounded convergence theorem stated in Section 2.5, we can let $n \to \infty$ in (5), obtaining

$$\lim_{n \to \infty} P(X_n = y) = \sum_x \pi_0(x) \pi(y).$$

Since $\sum_x \pi_0(x) = 1$, we conclude that

(6) $$\lim_{n \to \infty} P(X_n = y) = \pi(y), \qquad y \in \mathscr{S}.$$

Formula (6) states that, regardless of the initial distribution, for large values of n the distribution of X_n is approximately equal to the stationary distribution π. It implies that π is the unique stationary distribution. For if there were some other stationary distribution we could use it for the initial distribution π_0. From (4) and (6) we would conclude that $\pi_0(y) = \pi(y)$, $y \in \mathscr{S}$.

Consider a system described by a Markov chain having transition function P and unique stationary distribution π. Suppose we start observing the system after it has been going on for some time, say n_0 units of time for some large positive integer n_0. In effect, we observe Y_n, $n \geq 0$, where

$$Y_n = X_{n+n_0}, \qquad n \geq 0.$$

The random variables Y_n, $n \geq 0$, also form a Markov chain with transition function P. In order to determine unique probabilities for events defined in terms of the Y_n chain, we need to know its initial distribution, which is the same as the distribution of X_{n_0}. In most practical applications it is very

2.2. Examples

hard to determine this distribution exactly. We may have no choice but to assume that Y_n, $n \geq 0$, has the stationary distribution π for its initial distribution. This is a reasonable assumption if (2) holds and n_0 is large.

2.2. Examples

In this section we will consider some examples in which we can show directly that a unique stationary distribution exists and find simple formulas for it.

In Section 1.1 we discussed the two-state Markov chain on $\mathscr{S} = \{0, 1\}$ having transition matrix

$$\begin{array}{c c} & \begin{array}{cc} 0 & 1 \end{array} \\ \begin{array}{c} 0 \\ 1 \end{array} & \left[\begin{array}{cc} 1-p & p \\ q & 1-q \end{array} \right]. \end{array}$$

We saw that if $p + q > 0$, the chain has a unique stationary distribution π, determined by

$$\pi(0) = \frac{q}{p+q} \quad \text{and} \quad \pi(1) = \frac{p}{p+q}.$$

We also saw that if $0 < p + q < 2$, then (2) holds.

For Markov chains having a finite number of states, stationary distributions can be found by solving a finite system of linear equations.

Example 1. Consider a Markov chain having state space $\mathscr{S} = \{0, 1, 2\}$ and transition matrix

$$\begin{array}{c c} & \begin{array}{ccc} 0 & 1 & 2 \end{array} \\ \begin{array}{c} 0 \\ 1 \\ 2 \end{array} & \left[\begin{array}{ccc} \frac{1}{3} & \frac{1}{3} & \frac{1}{3} \\ \frac{1}{4} & \frac{1}{2} & \frac{1}{4} \\ \frac{1}{6} & \frac{1}{3} & \frac{1}{2} \end{array} \right]. \end{array}$$

Show that this chain has a unique stationary distribution π and find π.

Formula (1) in this case gives us the three equations

$$\frac{\pi(0)}{3} + \frac{\pi(1)}{4} + \frac{\pi(2)}{6} = \pi(0),$$

$$\frac{\pi(0)}{3} + \frac{\pi(1)}{2} + \frac{\pi(2)}{3} = \pi(1),$$

$$\frac{\pi(0)}{3} + \frac{\pi(1)}{4} + \frac{\pi(2)}{2} = \pi(2).$$

$\sum_x \pi(x) = 1$ gives us the fourth equation

$$\pi(0) + \pi(1) + \pi(2) = 1.$$

By subtracting twice the first equation from the second equation, we eliminate the term involving $\pi(2)$ and find that $\pi(1) = 5\pi(0)/3$. We conclude from the first equation that $\pi(2) = 3\pi(0)/2$. From the fourth equation we now see that

$$\pi(0)(1 + \tfrac{5}{3} + \tfrac{3}{2}) = 1,$$

and hence that

$$\pi(0) = \tfrac{6}{25}.$$

Thus

$$\pi(1) = \tfrac{5}{3} \cdot \tfrac{6}{25} = \tfrac{2}{5}$$

and

$$\pi(2) = \tfrac{3}{2} \cdot \tfrac{6}{25} = \tfrac{9}{25}.$$

It is readily seen that these numbers satisfy all four equations. Since they are nonnegative, the unique stationary distribution is given by

$$\pi(0) = \tfrac{6}{25}, \qquad \pi(1) = \tfrac{2}{5}, \qquad \text{and} \qquad \pi(2) = \tfrac{9}{25}.$$

Though it is not easy to see directly, (2) holds for this chain (see Section 2.7).

2.2.1. Birth and death chain. Consider a birth and death chain on $\{0, 1, \ldots, d\}$ or on the nonnegative integers. In the latter case we set $d = \infty$. We assume without further mention that the chain is irreducible, i.e., that

$$p_x > 0 \quad \text{for} \quad 0 \le x < d$$

and

$$q_x > 0 \quad \text{for} \quad 0 < x \le d$$

if d is finite, and that

$$p_x > 0 \quad \text{for} \quad 0 \le x < \infty$$

and

$$q_x > 0 \quad \text{for} \quad 0 < x < \infty$$

if d is infinite.

Suppose d is infinite. The system of equations

$$\sum_x \pi(x) P(x, y) = \pi(y), \qquad y \in \mathscr{S},$$

2.2. Examples

becomes
$$\pi(0)r_0 + \pi(1)q_1 = \pi(0),$$
$$\pi(y-1)p_{y-1} + \pi(y)r_y + \pi(y+1)q_{y+1} = \pi(y), \quad y \geq 1.$$

Since
$$p_y + q_y + r_y = 1,$$
these equations reduce to

(7)
$$q_1\pi(1) - p_0\pi(0) = 0,$$
$$q_{y+1}\pi(y+1) - p_y\pi(y) = q_y\pi(y) - p_{y-1}\pi(y-1), \quad y \geq 1.$$

It follows easily from (7) and induction that
$$q_{y+1}\pi(y+1) - p_y\pi(y) = 0, \quad y \geq 0,$$
and hence that
$$\pi(y+1) = \frac{p_y}{q_{y+1}} \pi(y), \quad y \geq 0.$$

Consequently,

(8) $$\pi(x) = \frac{p_0 \cdots p_{x-1}}{q_1 \cdots q_x} \pi(0), \quad x \geq 1.$$

Set

(9) $$\pi_x = \begin{cases} 1, & x = 0, \\ \dfrac{p_0 \cdots p_{x-1}}{q_1 \cdots q_x}, & x \geq 1. \end{cases}$$

Then (8) can be written as

(10) $$\pi(x) = \pi_x \pi(0), \quad x \geq 0.$$

Conversely, (1) follows from (10).

Suppose now that $\sum_x \pi_x < \infty$ or, equivalently, that

(11) $$\sum_{x=1}^{\infty} \frac{p_0 \cdots p_{x-1}}{q_1 \cdots q_x} < \infty.$$

We conclude from (10) that the birth and death chain has a unique stationary distribution, given by

(12) $$\pi(x) = \frac{\pi_x}{\sum_{y=0}^{\infty} \pi_y}, \quad x \geq 0.$$

Suppose instead that (11) fails to hold, i.e., that

(13) $$\sum_{x=0}^{\infty} \pi_x = \infty.$$

We conclude from (10) and (13) that any solution to (1) is either identically zero or has infinite sum, and hence that there is no stationary distribution.

In summary, we see that the chain has a stationary distribution if and only if (11) holds, and that the stationary distribution, when it exists, is given by (9) and (12).

Suppose now that $d < \infty$. By essentially the same arguments used to obtain (12), we conclude that the unique stationary distribution is given by

$$\pi(x) = \frac{\pi_x}{\sum_{y=0}^{d} \pi_y}, \qquad 0 \le x \le d, \tag{14}$$

where π_x, $0 \le x \le d$, is given by (9).

Example 2. Consider the Ehrenfest chain introduced in Section 1.3 and suppose that $d = 3$. Find the stationary distribution.

The transition matrix of the chain is

$$\begin{array}{c} \\ 0 \\ 1 \\ 2 \\ 3 \end{array} \begin{array}{cccc} 0 & 1 & 2 & 3 \\ \left[\begin{array}{cccc} 0 & 1 & 0 & 0 \\ \frac{1}{3} & 0 & \frac{2}{3} & 0 \\ 0 & \frac{2}{3} & 0 & \frac{1}{3} \\ 0 & 0 & 1 & 0 \end{array}\right] \end{array}.$$

This is an irreducible birth and death chain in which $\pi_0 = 1$,

$$\pi_1 = \frac{1}{\frac{1}{3}} = 3, \qquad \pi_2 = \frac{1 \cdot \frac{2}{3}}{\frac{1}{3} \cdot \frac{2}{3}} = 3,$$

and

$$\pi_3 = \frac{1 \cdot \frac{2}{3} \cdot \frac{1}{3}}{\frac{1}{3} \cdot \frac{2}{3} \cdot 1} = 1.$$

Thus the unique stationary distribution is given by

$$\pi(0) = \tfrac{1}{8}, \qquad \pi(1) = \tfrac{3}{8}, \qquad \pi(2) = \tfrac{3}{8}, \qquad \text{and} \qquad \pi(3) = \tfrac{1}{8}.$$

Formula (2) does not hold for the chain in Example 2 since $P^n(x, x) = 0$ for odd values of n. We can modify the Ehrenfest chain slightly and avoid such "periodic" behavior.

Example 3. Modified Ehrenfest chain. Suppose we have two boxes labeled 1 and 2 and d balls labeled $1, 2, \ldots, d$. Initially some of the balls are in box 1 and the remainder are in box 2. An integer is selected at random from $1, 2, \ldots, d$, and the ball labeled by that integer is removed from its box. We now select at random one of the two boxes and put the removed ball into this box. The procedure is repeated indefinitely, the selections being made independently. Let X_n denote the number of balls in box 1 after the nth trial. Then X_n, $n \geq 0$, is a Markov chain on $\mathscr{S} = \{0, 1, \ldots, d\}$. Find the stationary distribution of the chain for $d = 3$.

The transition matrix of this chain, for $d = 3$, is

$$\begin{array}{c} \\ 0 \\ 1 \\ 2 \\ 3 \end{array} \begin{array}{cccc} 0 & 1 & 2 & 3 \end{array} \\ \left[\begin{array}{cccc} \frac{1}{2} & \frac{1}{2} & 0 & 0 \\ \frac{1}{6} & \frac{1}{2} & \frac{1}{3} & 0 \\ 0 & \frac{1}{3} & \frac{1}{2} & \frac{1}{6} \\ 0 & 0 & \frac{1}{2} & \frac{1}{2} \end{array} \right].$$

To see why P is given as indicated, we will compute $P(1, y)$, $0 \leq y \leq 3$. We start with one ball in box 1 and two balls in box 2. Thus $P(1, 0)$ is the probability that the ball selected is from box 1 and the box selected is box 2. Thus

$$P(1, 0) = \tfrac{1}{3} \cdot \tfrac{1}{2} = \tfrac{1}{6}.$$

Secondly, $P(1, 2)$ is the probability that the ball selected is from box 2 and the box selected is box 1. Thus

$$P(1, 2) = \tfrac{2}{3} \cdot \tfrac{1}{2} = \tfrac{1}{3}.$$

Clearly $P(1, 3) = 0$, since at most one ball is transferred at a time. Finally, $P(1, 1)$ can be obtained by subtracting $P(1, 0) + P(1, 2) + P(1, 3)$ from 1. Alternatively, $P(1, 1)$ is the probability that either the selected ball is from box 1 and the selected box is box 1 or the selected ball is from box 2 and the selected box is box 2. Thus

$$P(1, 1) = \tfrac{1}{3} \cdot \tfrac{1}{2} + \tfrac{2}{3} \cdot \tfrac{1}{2} = \tfrac{1}{2}.$$

The other probabilities are computed similarly. This Markov chain is an irreducible birth and death chain. It is easily seen that π_x, $0 \leq x \leq 3$, are the same as in the previous example and hence that the stationary distribution is again given by

$$\pi(0) = \tfrac{1}{8}, \quad \pi(1) = \tfrac{3}{8}, \quad \pi(2) = \tfrac{3}{8}, \quad \text{and} \quad \pi(3) = \tfrac{1}{8}.$$

It follows from the results in Section 2.7 that (2) holds for the chain in Example 3.

2.2.2. Particles in a box.

A Markov chain that arises in several applied contexts can be described as follows. Suppose that ξ_n particles are added to a box at times $n = 1, 2, \ldots$, where ξ_n, $n \geq 1$, are independent and have a Poisson distribution with common parameter λ. Suppose that each particle in the box at time n, independently of all the other particles in the box and independently of how particles are added to the box, has probability $p < 1$ of remaining in the box at time $n + 1$ and probability $q = 1 - p$ of being removed from the box at time $n + 1$. Let X_n denote the number of particles in the box at time n. Then X_n, $n \geq 0$, is a Markov chain. We will find the stationary distribution of this chain. We will also find an explicit formula for $P^n(x, y)$ and use this formula to show directly that (2) holds.

The same Markov chain can be used to describe a telephone exchange, where ξ_n is the number of new calls starting at time n, q is the probability that a call in progress at time n terminates by time $n + 1$, and X_n is the number of calls in progress at time n.

We will now analyze this Markov chain. Let $R(X_n)$ denote the number of particles present at time n that remain in the box at time $n + 1$. Then

$$X_{n+1} = \xi_{n+1} + R(X_n).$$

Clearly

$$P(R(X_n) = z \mid X_n = x) = \binom{x}{z} p^z (1-p)^{x-z}, \qquad 0 \leq z \leq x,$$

and

$$P(\xi_n = z) = \frac{\lambda^z e^{-\lambda}}{z!}, \qquad z \geq 0.$$

Since

$$P(X_{n+1} = y \mid X_n = x) = \sum_{z=0}^{\min(x,y)} P(R(X_n) = z, \xi_{n+1} = y - z \mid X_n = x)$$

$$= \sum_{z=0}^{\min(x,y)} P(\xi_{n+1} = y - z) P(R(X_n) = z \mid X_n = x),$$

we conclude that

(15) $$P(x, y) = \sum_{z=0}^{\min(x,y)} \frac{\lambda^{y-z} e^{-\lambda}}{(y-z)!} \binom{x}{z} p^z (1-p)^{x-z}.$$

It follows from (15) or from the original description of the process that $P(x, y) > 0$ for all $x \geq 0$ and $y \geq 0$, and hence that the chain is irreducible.

Suppose X_n has a Poisson distribution with parameter t. Then $R(X_n)$ has a Poisson distribution with parameter pt. For

$$P(R(X_n) = y) = \sum_{x=y}^{\infty} P(X_n = x, R(X_n) = y)$$

$$= \sum_{x=y}^{\infty} P(X_n = x) P(R(X_n) = y \mid X_n = x)$$

$$= \sum_{x=y}^{\infty} \frac{t^x e^{-t}}{x!} \binom{x}{y} p^y (1-p)^{x-y}$$

2.2 Examples

$$= \sum_{x=y}^{\infty} \frac{t^x e^{-t}}{y!(x-y)!} p^y(1-p)^{x-y}$$

$$= \frac{(pt)^y e^{-t}}{y!} \sum_{x=y}^{\infty} \frac{(t(1-p))^{x-y}}{(x-y)!}$$

$$= \frac{(pt)^y e^{-t}}{y!} \sum_{z=0}^{\infty} \frac{(t(1-p))^z}{z!}$$

$$= \frac{(pt)^y e^{-t}}{y!} e^{t(1-p)}$$

$$= \frac{(pt)^y e^{-pt}}{y!},$$

which shows that $R(X_n)$ has the indicated Poisson distribution.

We will now show that the stationary distribution is Poisson with parameter t for suitable t. Let X_0 have such a distribution. Then $X_1 = \xi_1 + R(X_0)$ is the sum of independent random variables having Poisson distributions with parameters λ and pt respectively. Thus X_1 has a Poisson distribution with parameter $\lambda + pt$. The distribution of X_1 will agree with that of X_0 if $t = \lambda + pt$, i.e., if

$$t = \frac{\lambda}{1-p} = \frac{\lambda}{q}.$$

We conclude that the Markov chain has a stationary distribution π which is a Poisson distribution with parameter λ/q, i.e., such that

(16) $$\pi(x) = \frac{(\lambda/q)^x e^{-\lambda/q}}{x!}, \quad x \geq 0.$$

Finally we will derive a formula for $P^n(x, y)$. Suppose X_0 has a Poisson distribution with parameter t. It is left as an exercise for the reader to show that X_n has a Poisson distribution with parameter

$$tp^n + \frac{\lambda}{q}(1-p^n).$$

Thus

$$\sum_{x=0}^{\infty} \frac{e^{-t}t^x}{x!} P^n(x,y) = P(X_n = y)$$

$$= \exp\left[-\left(tp^n + \frac{\lambda}{q}(1-p^n)\right)\right] \frac{\left[tp^n + \frac{\lambda}{q}(1-p^n)\right]^y}{y!},$$

and hence

(17) $$\sum_{x=0}^{\infty} t^x \frac{P^n(x,y)}{x!} = e^{-\lambda(1-p^n)/q} e^{t(1-p^n)} \frac{\left[tp^n + \frac{\lambda}{q}(1-p^n)\right]^y}{y!}.$$

Now if
$$\sum_{x=0}^{\infty} c_x t^x = \left(\sum_{x=0}^{\infty} b_x t^x\right)\left(\sum_{x=0}^{\infty} a_x t^x\right),$$
where each power series has a positive radius of convergence, then
$$c_x = \sum_{z=0}^{x} a_z b_{x-z}.$$
If $a_z = 0$ for $z > y$, then
$$c_x = \sum_{z=0}^{\min(x,y)} a_z b_{x-z}.$$
Using this with (17) and the binomial expansion, we conclude that
$$P^n(x, y) = \frac{x!\, e^{-\lambda(1-p^n)/q}}{y!} \sum_{z=0}^{\min(x,y)} \binom{y}{z} p^{nz} \left[\frac{\lambda}{q}(1 - p^n)\right]^{y-z} \frac{(1 - p^n)^{x-z}}{(x - z)!},$$
which simplifies slightly to

(18) $$P^n(x, y) = e^{-\lambda(1-p^n)/q} \sum_{z=0}^{\min(x,y)} \binom{x}{z} p^{nz}(1 - p^n)^{x-z} \frac{\left[\frac{\lambda}{q}(1 - p^n)\right]^{y-z}}{(y - z)!}.$$

Since $0 \le p < 1$,
$$\lim_{n \to \infty} p^n = 0.$$
Thus as $n \to \infty$, the terms in the sum in (18) all approach zero except for the term corresponding to $z = 0$. We conclude that

(19) $$\lim_{n \to \infty} P^n(x, y) = \frac{e^{-\lambda/q} \left(\frac{\lambda}{q}\right)^y}{y!} = \pi(y), \qquad x, y \ge 0.$$

Thus (2) holds for this chain, and consequently the distribution π given by (16) is the unique stationary distribution of the chain.

2.3. Average number of visits to a recurrent state

Consider an irreducible birth and death chain with stationary distribution π. Suppose that $P(x, x) = r_x = 0$, $x \in \mathcal{S}$, as in the Ehrenfest chain and the gambler's ruin chain. Then at each transition the birth and death chain moves either one step to the right or one step to the left. Thus the chain can return to its starting point only after an even number of transitions. In other words, $P^n(x, x) = 0$ for odd values of n. For such a chain the formula
$$\lim_{n \to \infty} P^n(x, y) = \pi(y), \qquad y \in \mathcal{S},$$
clearly fails to hold.

2.3. Average number of visits to a recurrent state

There is a way to handle such situations. Let a_n, $n \geq 0$, be a sequence of numbers. If

(20) $$\lim_{n \to \infty} a_n = L$$

for some finite number L, then

(21) $$\lim_{n \to \infty} \frac{1}{n} \sum_{m=1}^{n} a_m = L.$$

Formula (21) can hold, however, even if (20) fails to hold. For example, if $a_n = 0$ for n odd and $a_n = 1$ for n even, then a_n has no limit as $n \to \infty$, but

$$\lim_{n \to \infty} \frac{1}{n} \sum_{m=1}^{n} a_m = \frac{1}{2}.$$

In this section we will show that

$$\lim_{n \to \infty} \frac{1}{n} \sum_{m=1}^{n} P^m(x, y)$$

exists for every pair x, y of states for an arbitrary Markov chain. In Section 2.5 we will use the existence of these limits to determine which Markov chains have stationary distributions and when there is such a unique distribution.

Recall that

$$1_y(z) = \begin{cases} 1, & z = y, \\ 0, & z \neq y, \end{cases}$$

and that

(22) $$E_x(1_y(X_n)) = P_x(X_n = y) = P^n(x, y).$$

Set

$$N_n(y) = \sum_{m=1}^{n} 1_y(X_m)$$

and

$$G_n(x, y) = \sum_{m=1}^{n} P^m(x, y).$$

Then $N_n(y)$ denotes the number of visits of the Markov chain to y during times $m = 1, \ldots, n$. The expected number of such visits for a chain starting at x is given according to (22) by

(23) $$E_x(N_n(y)) = G_n(x, y).$$

Let y be a transient state. Then

$$\lim_{n \to \infty} N_n(y) = N(y) < \infty \quad \text{with probability one,}$$

and

$$\lim_{n\to\infty} G_n(x, y) = G(x, y) < \infty, \qquad x \in \mathscr{S}.$$

It follows that

(24) $$\lim_{n\to\infty} \frac{N_n(y)}{n} = 0 \quad \text{with probability one,}$$

and that

(25) $$\lim_{n\to\infty} \frac{G_n(x, y)}{n} = 0, \qquad x \in \mathscr{S}.$$

Observe that $N_n(y)/n$ is the proportion of the first n units of time that the chain is in state y and that $G_n(x, y)/n$ is the expected value of this proportion for a chain starting at x.

Suppose now that y is a recurrent state. Let $m_y = E_y(T_y)$ denote the mean return time to y for a chain starting at y if this return time has finite expectation, and set $m_y = \infty$ otherwise. Let $1_{\{T_y < \infty\}}$ denote the random variable that is 1 if $T_y < \infty$ and 0 if $T_y = \infty$.

We will use the strong law of large numbers to prove the main result of this section, namely, Theorem 1 below.

Strong Law of Large Numbers. *Let $\xi_1, \xi_2, \ldots$ be independent identically distributed random variables. If these random variables have finite mean μ, then*

$$\lim_{n\to\infty} \frac{\xi_1 + \cdots + \xi_n}{n} = \mu \quad \text{with probability one.}$$

If these random variables are nonnegative and fail to have finite expectation, then this limit holds, provided that we set $\mu = +\infty$.

This important theorem is proved in advanced probability texts.

Theorem 1 *Let y be a recurrent state. Then*

(26) $$\lim_{n\to\infty} \frac{N_n(y)}{n} = \frac{1_{\{T_y < \infty\}}}{m_y} \quad \text{with probability one,}$$

and

(27) $$\lim_{n\to\infty} \frac{G_n(x, y)}{n} = \frac{\rho_{xy}}{m_y}, \qquad x \in \mathscr{S}.$$

These formulas are intuitively very reasonable. Once a chain reaches y, it returns to y "on the average every m_y units of time." Thus if $T_y < \infty$ and n is large, the proportion of the first n units of time that the chain is in

2.3. Average number of visits to a recurrent state

state y should be about $1/m_y$. Formula (27) should follow from (26) by taking expectations.

From Corollary 1 of Chapter 1 and the above theorem, we immediately obtain the next result.

Corollary 1 *Let C be an irreducible closed set of recurrent states. Then*

(28) $$\lim_{n \to \infty} \frac{G_n(x, y)}{n} = \frac{1}{m_y}, \quad x, y \in C,$$

and if $P(X_0 \in C) = 1$, then with probability one

(29) $$\lim_{n \to \infty} \frac{N_n(y)}{n} = \frac{1}{m_y}, \quad y \in C.$$

If $m_y = \infty$ the right sides of (26)–(29) all equal zero, and hence (24) and (25) hold.

Proof. In order to verify Theorem 1, we need to introduce some additional random variables. Consider a Markov chain starting at a recurrent state y. With probability one it returns to y infinitely many times. For $r \geq 1$ let T_y^r denote the time of the rth visit to y, so that

$$T_y^r = \min(n \geq 1 : N_n(y) = r).$$

Set $W_y^1 = T_y^1 = T_y$ and for $r \geq 2$ let $W_y^r = T_y^r - T_y^{r-1}$ denote the waiting time between the $(r-1)$th visit to y and the rth visit to y. Clearly

$$T_y^r = W_y^1 + \cdots + W_y^r.$$

The random variables $W_y^1, W_y^2, \ldots$ are independent and identically distributed and hence they have common mean $E_y(W_y^1) = E_y(T_y) = m_y$. This result should be intuitively obvious, since every time the chain returns to y it behaves from then on just as would a chain starting out initially at y. One can give a rigorous proof of this result by using (27) of Chapter 1 to show that for $r \geq 1$

$$P(W_y^{r+1} = m_{r+1} \mid W_y^1 = m_1, \ldots, W_y^r = m_r) = P_y(W_y^1 = m_{r+1});$$

and then showing by induction that

$$P_y(W_y^1 = m_1, \ldots, W_y^r = m_r) = P_y(W_y^1 = m_1) \cdots P_y(W_y^1 = m_r).$$

The strong law of large numbers implies that

$$\lim_{k \to \infty} \frac{W_y^1 + W_y^2 + \cdots + W_y^k}{k} = m_y \quad \text{with probability one,}$$

i.e., that

(30) $$\lim_{k \to \infty} \frac{T_y^k}{k} = m_y \quad \text{with probability one.}$$

Set $r = N_n(y)$. By time n the chain has made exactly r visits to y. Thus the rth visit to y occurs on or before time n, and the $(r+1)$th visit to y occurs after time n; that is,

$$T_y^{N_n(y)} \leq n < T_y^{N_n(y)+1},$$

and hence

$$\frac{T_y^{N_n(y)}}{N_n(y)} \leq \frac{n}{N_n(y)} \leq \frac{T_y^{N_n(y)+1}}{N_n(y)},$$

or at least these results hold for n large enough so that $N_n(y) \geq 1$. Since $N_n(y) \to \infty$ with probability one as $n \to \infty$, these inequalities and (30) together imply that

$$\lim_{n \to \infty} \frac{n}{N_n(y)} = m_y \quad \text{with probability one,}$$

or, equivalently, that (29) holds.

Let y be a recurrent state as before, but let X_0 have an arbitrary distribution. Then the chain may never reach y. If it does reach y, however, the above argument is valid; and hence, with probability one, $N_n(y)/n \to 1\{T_y < \infty\}/m_y$ as $n \to \infty$. Thus (26) is valid.

By definition $0 \leq N_n(y) \leq n$, and hence

(31) $$0 \leq \frac{N_n(y)}{n} \leq 1.$$

A theorem from measure theory, known as the dominated convergence theorem, allows us to conclude from (26) and (31) that

$$\lim_{n \to \infty} E_x \left(\frac{N_n(y)}{n} \right) = E_x \left(\frac{1_{\{T_y < \infty\}}}{m_y} \right) = \frac{P_x(T_y < \infty)}{m_y} = \frac{\rho_{xy}}{m_y}$$

and hence from (23) that (27) holds. This completes the proof of Theorem 1. ∎

2.4. Null recurrent and positive recurrent states

A recurrent state y is called *null recurrent* if $m_y = \infty$. From Theorem 1 we see that if y is null recurrent, then

(32) $$\lim_{n \to \infty} \frac{G_n(x, y)}{n} = \lim_{n \to \infty} \frac{\sum_{m=1}^{n} P^m(x, y)}{n} = 0, \quad x \in \mathcal{S}.$$

2.4. Null recurrent and positive recurrent states

(It can be shown that if y is null recurrent, then

(33) $$\lim_{n \to \infty} P^n(x, y) = 0, \qquad x \in \mathscr{S},$$

which is a stronger result than (32). We will not prove (33), since it will not be needed later and its proof is rather difficult.)

A recurrent state y is called *positive recurrent* if $m_y < \infty$. It follows from Theorem 1 that if y is positive recurrent, then

$$\lim_{n \to \infty} \frac{G_n(y, y)}{n} = \frac{1}{m_y} > 0.$$

Thus (32) and (33) fail to hold for positive recurrent states.

Consider a Markov chain starting out in a recurrent state y. It follows from Theorem 1 that if y is null recurrent, then, with probability one, the proportion of time the chain is in state y during the first n units of time approaches zero as $n \to \infty$. On the other hand, if y is a positive recurrent state, then, with probability one, the proportion of time the chain is in state y during the first n units of time approaches the positive limit $1/m_y$ as $n \to \infty$.

The next result is closely related to Theorem 2 of Chapter 1.

Theorem 2 *Let x be a positive recurrent state and suppose that x leads to y. Then y is positive recurrent.*

Proof. It follows from Theorem 2 of Chapter 1 that y leads to x. Thus there exist positive integers n_1 and n_2 such that

$$P^{n_1}(y, x) > 0 \quad \text{and} \quad P^{n_2}(x, y) > 0.$$

Now

$$P^{n_1+m+n_2}(y, y) \geq P^{n_1}(y, x)P^m(x, x)P^{n_2}(x, y),$$

and by summing on $m = 1, 2, \ldots, n$ and dividing by n, we conclude that

$$\frac{G_{n_1+n+n_2}(y, y)}{n} - \frac{G_{n_1+n_2}(y, y)}{n} \geq P^{n_1}(y, x)P^{n_2}(x, y) \frac{G_n(x, x)}{n}.$$

As $n \to \infty$, the left side of this inequality converges to $1/m_y$ and the right side converges to

$$\frac{P^{n_1}(y, x)P^{n_2}(x, y)}{m_x}.$$

Hence

$$\frac{1}{m_y} \geq \frac{P^{n_1}(y, x)P^{n_2}(x, y)}{m_x} > 0,$$

and consequently $m_y < \infty$. This shows that y is positive recurrent. ∎

From this theorem and from Theorem 2 of Chapter 1 we see that if C is an irreducible closed set, then every state in C is transient, every state in C is null recurrent, or every state in C is positive recurrent. A Markov chain is called a *null recurrent chain* if all its states are null recurrent and a *positive recurrent chain* if all its states are positive recurrent. We see therefore that an irreducible Markov chain is a transient chain, a null recurrent chain, or a positive recurrent chain.

If C is a finite closed set of states, then C has at least one positive recurrent state. For

$$\sum_{y \in C} P^m(x, y) = 1, \quad x \in C,$$

and by summing on $m = 1, \ldots, n$ and dividing by n we find that

$$\sum_{y \in C} \frac{G_n(x, y)}{n} = 1, \quad x \in C.$$

If C is finite and each state in C is transient or null recurrent, then (25) holds and hence

$$1 = \lim_{n \to \infty} \sum_{y \in C} \frac{G_n(x, y)}{n}$$

$$= \sum_{y \in C} \lim_{n \to \infty} \frac{G_n(x, y)}{n} = 0,$$

a contradiction.

We are now able to sharpen Theorem 3 of Chapter 1.

Theorem 3 *Let C be a finite irreducible closed set of states. Then every state in C is positive recurrent.*

Proof. The proof of this theorem is now almost immediate. Since C is a finite closed set, there is at least one positive recurrent state in C. Since C is irreducible, every state in C is positive recurrent by Theorem 2. ∎

Corollary 2 *An irreducible Markov chain having a finite number of states is positive recurrent.*

Corollary 3 *A Markov chain having a finite number of states has no null recurrent states.*

Proof. Corollary 2 follows immediately from Theorem 3. To verify Corollary 3, observe that if y is a recurrent state, then, by Theorem 4 of Chapter 1, y is contained in an irreducible closed set C of recurrent states. Since C is necessarily finite, it follows from Theorem 3 that all states in C, including y itself, are positive recurrent. Thus every recurrent state is positive recurrent, and hence there are no null recurrent states. ∎

2.5. Existence and uniqueness of stationary distributions

Example 4. Consider the Markov chain described in Example 10 of Chapter 1. We have seen that 1 and 2 are transient states and that 0, 3, 4, and 5 are recurrent states. We now see that these recurrent states are necessarily positive recurrent.

2.5. Existence and uniqueness of stationary distributions

In this section we will determine which Markov chains have stationary distributions and when there is a unique such distribution. In our discussion we will need to interchange summations and limits on several occasions. This is justified by the following standard elementary result in analysis, which we state without proof.

Bounded Convergence Theorem. *Let $a(x)$, $x \in \mathscr{S}$, be nonnegative numbers having finite sum, and let $b_n(x)$, $x \in \mathscr{S}$ and $n \geq 1$, be such that $|b_n(x)| \leq 1$, $x \in \mathscr{S}$ and $n \geq 1$, and*

$$\lim_{n \to \infty} b_n(x) = b(x), \qquad x \in \mathscr{S}.$$

Then

$$\lim_{n \to \infty} \sum_x a(x) b_n(x) = \sum_x a(x) b(x).$$

Let π be a stationary distribution and let m be a positive integer. Then by (3)

$$\sum_z \pi(z) P^m(z, x) = \pi(x).$$

Summing this equation on $m = 1, 2, \ldots, n$ and dividing by n, we conclude that

$$(34) \qquad \sum_z \pi(z) \frac{G_n(z, x)}{n} = \pi(x), \qquad x \in \mathscr{S}.$$

Theorem 4 *Let π be a stationary distribution. If x is a transient state or a null recurrent state, then $\pi(x) = 0$.*

Proof. If x is a transient state or a null recurrent state,

$$(35) \qquad \lim_{n \to \infty} \frac{G_n(z, x)}{n} = 0, \qquad x \in \mathscr{S},$$

as shown in Sections 2.3 and 2.4. It follows from (34), (35), and the bounded convergence theorem that

$$\pi(x) = \lim_{n \to \infty} \sum_z \pi(z) \frac{G_n(z, x)}{n} = 0,$$

as desired. ∎

It follows from this theorem that a Markov chain with no positive recurrent states does not have a stationary distribution.

Theorem 5 *An irreducible positive recurrent Markov chain has a unique stationary distribution π, given by*

$$(36) \qquad \pi(x) = \frac{1}{m_x}, \qquad x \in \mathcal{S}.$$

Proof. It follows from Theorem 1 and the assumptions of this theorem that

$$(37) \qquad \lim_{n \to \infty} \frac{G_n(z, x)}{n} = \frac{1}{m_x}, \qquad x, z \in \mathcal{S}.$$

Suppose π is a stationary distribution. We see from (34), (37), and the bounded convergence theorem that

$$\pi(x) = \lim_{n \to \infty} \sum_z \pi(z) \frac{G_n(z, x)}{n}$$

$$= \frac{1}{m_x} \sum_z \pi(z) = \frac{1}{m_x}.$$

Thus if there is a stationary distribution, it must be given by (36).

To complete the proof of the theorem we need to show that the function $\pi(x)$, $x \in \mathcal{S}$, defined by (36) is indeed a stationary distribution. It is clearly nonnegative, so we need only show that

$$(38) \qquad \sum_x \frac{1}{m_x} = 1$$

and

$$(39) \qquad \sum_x \frac{1}{m_x} P(x, y) = \frac{1}{m_y}, \qquad y \in \mathcal{S}.$$

Toward this end we observe first that

$$\sum_x P^m(z, x) = 1.$$

Summing on $m = 1, \ldots, n$ and dividing by n, we conclude that

$$(40) \qquad \sum_x \frac{G_n(z, x)}{n} = 1, \qquad z \in \mathcal{S}.$$

Next we observe that by (24) of Chapter 1

$$\sum_x P^m(z, x) P(x, y) = P^{m+1}(z, y).$$

2.5. Existence and uniqueness of stationary distributions

By again summing on $m = 1, \ldots, n$ and dividing by n, we conclude that

$$(41) \qquad \sum_x \frac{G_n(z, x)}{n} P(x, y) = \frac{G_{n+1}(z, y)}{n} - \frac{P(z, y)}{n}.$$

If $\mathscr{S}$ is finite, we conclude from (37) and (40) that

$$1 = \lim_{n \to \infty} \sum_x \frac{G_n(z, x)}{n} = \sum_x \frac{1}{m_x},$$

i.e., that (38) holds. Similarly, we conclude that (39) holds by letting $n \to \infty$ in (41). This completes the proof of the theorem if $\mathscr{S}$ is finite.

The argument to complete the proof for $\mathscr{S}$ infinite is more complicated, since we cannot directly interchange limits and sums as we did for $\mathscr{S}$ finite (the bounded convergence theorem is not applicable). Let $\mathscr{S}_1$ be a finite subset of $\mathscr{S}$. We see from (40) that

$$\sum_{x \in \mathscr{S}_1} \frac{G_n(z, x)}{n} \leq 1, \qquad z \in \mathscr{S}.$$

Since $\mathscr{S}_1$ is finite, we can let $n \to \infty$ in this inequality and conclude from (37) that

$$\sum_{x \in \mathscr{S}_1} \frac{1}{m_x} \leq 1.$$

The last inequality holds for any finite subset $\mathscr{S}_1$ of $\mathscr{S}$, and hence

$$(42) \qquad \sum_x \frac{1}{m_x} \leq 1.$$

For if the sum of $1/m_x$ over $x \in \mathscr{S}$ exceeded 1, the sum over some finite subset of $\mathscr{S}$ would also exceed 1.

Similarly, we conclude from (41) that if $\mathscr{S}_1$ is a finite subset of $\mathscr{S}$, then

$$\sum_{x \in \mathscr{S}_1} \frac{G_n(z, x)}{n} P(x, y) \leq \frac{G_{n+1}(z, y)}{n} - \frac{P(z, y)}{n}.$$

By letting $n \to \infty$ in this inequality and using (37), we obtain

$$\sum_{x \in \mathscr{S}_1} \frac{1}{m_x} P(x, y) \leq \frac{1}{m_y}.$$

We conclude, as in the proof of (42), that

$$(43) \qquad \sum_x \frac{1}{m_x} P(x, y) \leq \frac{1}{m_y}, \qquad y \in \mathscr{S}.$$

Next we will show that equality holds in (43). It follows from (42) that the sum on y of the right side of (43) is finite. If strict inequality held for some y, it would follow by summing (43) on y that

$$\sum_y \frac{1}{m_y} > \sum_y \left(\sum_x \frac{1}{m_x} P(x, y) \right)$$
$$= \sum_x \frac{1}{m_x} \left(\sum_y P(x, y) \right)$$
$$= \sum_x \frac{1}{m_x},$$

which is a contradiction. This proves that equality holds in (43), i.e., that (39) holds.

Set

$$c = \frac{1}{\sum_x \frac{1}{m_x}}.$$

Then by (39)

$$\pi(x) = \frac{c}{m_x}, \qquad x \in \mathscr{S},$$

defines a stationary distribution. Thus by the first part of the proof of this theorem

$$\frac{c}{m_x} = \frac{1}{m_x},$$

and hence $c = 1$. This proves that (38) holds and completes the proof of the theorem. ∎

From Theorems 4 and 5 we immediately obtain

Corollary 4 *An irreducible Markov chain is positive recurrent if and only if it has a stationary distribution.*

Example 5. Consider an irreducible birth and death chain on the nonnegative integers. Find necessary and sufficient conditions for the chain to be
(a) positive recurrent,
(b) null recurrent,
(c) transient.

From Section 2.2.1 we see that the chain has a stationary distribution if and only if

(44) $$\sum_{x=1}^{\infty} \frac{p_0 \cdots p_{x-1}}{q_1 \cdots q_x} < \infty.$$

2.5. Existence and uniqueness of stationary distributions

Thus (44) is necessary and sufficient for the chain to be positive recurrent. We saw in Section 1.7 that

$$(45) \qquad \sum_{x=1}^{\infty} \frac{q_1 \cdots q_x}{p_1 \cdots p_x} < \infty$$

is a necessary and sufficient condition for the chain to be transient. For the chain to be null recurrent, it is necessary and sufficient that (44) and (45) both fail to hold. Thus

$$(46) \qquad \sum_{x=1}^{\infty} \frac{q_1 \cdots q_x}{p_1 \cdots p_x} = \infty \quad \text{and} \quad \sum_{x=1}^{\infty} \frac{p_0 \cdots p_{x-1}}{q_1 \cdots q_x} = \infty$$

are necessary and sufficient conditions for the chain to be null recurrent. As an immediate consequence of Corollary 2 and Theorem 5 we obtain

Corollary 5 *If a Markov chain having a finite number of states is irreducible, it has a unique stationary distribution.*

Recall that $N_n(x)$ denotes the number of visits to x during times $m = 1, \ldots, n$. By combining Corollary 1 and Theorem 5 we get

Corollary 6 *Let X_n, $n \geq 0$, be an irreducible positive recurrent Markov chain having stationary distribution π. Then with probability one*

$$(47) \qquad \lim_{n \to \infty} \frac{N_n(x)}{n} = \pi(x), \qquad x \in \mathscr{S}.$$

2.5.1. Reducible chains.

Let π be a distribution on $\mathscr{S}$, i.e., let $\pi(x)$, $x \in \mathscr{S}$, be nonnegative numbers adding to one, and let C be a subset of $\mathscr{S}$. We say that π is *concentrated* on C if

$$\pi(x) = 0, \qquad x \notin C.$$

By essentially the same argument used to prove Theorem 5 we can obtain a somewhat more general result.

Theorem 6 *Let C be an irreducible closed set of positive recurrent states. Then the Markov chain has a unique stationary distribution π concentrated on C. It is given by*

$$(48) \qquad \pi(x) = \begin{cases} \dfrac{1}{m_x}, & x \in C, \\ 0, & \text{elsewhere.} \end{cases}$$

Suppose C_0 and C_1 are two distinct irreducible closed sets of positive recurrent states of a Markov chain. It follows from Theorem 6 that the Markov chain has a stationary distribution π_0 concentrated on C_0 and a different stationary distribution π_1 concentrated on C_1. Moreover, the distributions π_α defined for $0 \leq \alpha \leq 1$ by

$$\pi_\alpha(x) = (1 - \alpha)\pi_0(x) + \alpha\pi_1(x), \qquad x \in \mathscr{S},$$

are distinct stationary distributions (see Exercise 5).

By combining Theorems 4–6 and their consequences, we obtain

Corollary 7 *Let $\mathscr{S}_P$ denote the positive recurrent states of a Markov chain.*

(i) *If $\mathscr{S}_P$ is empty, the chain has no stationary distributions.*
(ii) *If $\mathscr{S}_P$ is a nonempty irreducible set, the chain has a unique stationary distribution.*
(iii) *If $\mathscr{S}_P$ is nonempty but not irreducible, the chain has an infinite number of distinct stationary distributions.*

Consider now a Markov chain having a finite number of states. Then every recurrent state is positive recurrent and there is at least one such state. There are two possibilities: either the set $\mathscr{S}_R$ of recurrent states is irreducible and there is a unique stationary distribution, or $\mathscr{S}_R$ can be decomposed into two or more irreducible closed sets and there is an infinite number of distinct stationary distributions. The latter possibility holds for a Markov chain on $\mathscr{S} = \{0, 1, \ldots, d\}$ in which $d > 0$ and 0 and d are both absorbing states. The gambler's ruin chain on $\{0, 1, \ldots, d\}$ and the genetics model in Example 7 of Chapter 1 are of this type. For such a chain any distribution π_α, $0 \leq \alpha \leq 1$, of the form

$$\pi_\alpha(x) = \begin{cases} 1 - \alpha, & x = 0, \\ \alpha, & x = d, \\ 0, & \text{elsewhere}, \end{cases}$$

is a stationary distribution.

Example 6. Consider the Markov chain introduced in Example 10 of Chapter 1. Find the stationary distribution concentrated on each of the irreducible closed sets.

We saw in Section 1.6 that the set of recurrent states for this chain is decomposed into the absorbing state 0 and the irreducible closed set $\{3, 4, 5\}$. Clearly the unique stationary distribution concentrated on $\{0\}$ is given by $\pi_0 = (1, 0, 0, 0, 0, 0)$. To find the unique stationary distribution concentrated on {3, 4, 5}, we must find nonnegative numbers $\pi(3)$, $\pi(4)$, and $\pi(5)$ summing to one and satisfying the three equations

$$\frac{\pi(3)}{6} + \frac{\pi(4)}{2} + \frac{\pi(5)}{4} = \pi(3)$$

$$\frac{\pi(3)}{3} = \pi(4)$$

$$\frac{\pi(3)}{2} + \frac{\pi(4)}{2} + \frac{3\pi(5)}{4} = \pi(5).$$

From the first two of these equations we find that $\pi(4) = \pi(3)/3$ and $\pi(5) = 8\pi(3)/3$. Thus

$$\pi(3)(1 + \tfrac{1}{3} + \tfrac{8}{3}) = 1,$$

from which we conclude that

$$\pi(3) = \tfrac{1}{4}, \qquad \pi(4) = \tfrac{1}{12}, \quad \text{and} \quad \pi(5) = \tfrac{2}{3}.$$

Consequently

$$\pi_1 = (0, 0, 0, \tfrac{1}{4}, \tfrac{1}{12}, \tfrac{2}{3})$$

is the stationary distribution concentrated on {3, 4, 5}.

2.6. Queuing chain

Consider the queuing chain introduced in Example 5 of Chapter 1. Recall that the number of customers arriving in unit time has density f and mean μ. Suppose that the chain is irreducible, which means that $f(0) > 0$ and $f(0) + f(1) < 1$ (see Exercise 37 of Chapter 1). In Chapter 1 we saw that the chain is recurrent if $\mu \leq 1$ and transient if $\mu > 1$. In Section 2.6.1 we will show that in the recurrent case

(49) $$m_0 = \frac{1}{1 - \mu}.$$

It follows from (49) that if $\mu < 1$, then $m_0 < \infty$ and hence 0 is a positive recurrent state. Thus by irreducibility the chain is positive recurrent. On the other hand, if $\mu = 1$, then $m_0 = \infty$ and hence 0 is a null recurrent state. We conclude that the queuing chain is null recurrent in this case. Therefore *an irreducible queuing chain is positive recurrent if $\mu < 1$ and null recurrent if $\mu = 1$, and transient if $\mu > 1$.*

*2.6.1. Proof. We will now verify (49). We suppose throughout the proof of this result that $f(0) > 0$, $f(0) + f(1) < 1$ and $\mu \leq 1$, so that the chain is irreducible and recurrent. Consider such a chain starting at the positive integer x. Then T_{x-1} denotes the time to go from state x to state $x - 1$, and $T_{y-1} - T_y$, $1 \leq y \leq x - 1$, denotes the time to go from state y to state $y - 1$. Since the queuing chain goes at most one step to the left at a time, the Markov property insures that the random variables

$$T_{x-1}, T_{x-2} - T_{x-1}, \ldots, T_0 - T_1$$

are independent. These random variables are identically distributed; for each of them is distributed as

$$\min(n > 0: \xi_1 + \cdots + \xi_n = n - 1),$$

i.e., as the smallest positive integer n such that the number of customers served by time n is one more than the number of new customers arriving by time n.

Let $G(t)$, $0 \leq t \leq 1$, denote the probability generation function of the time to go from state 1 to state 0. Then

$$(50) \qquad G(t) = \sum_{n=1}^{\infty} t^n P_1(T_0 = n).$$

The probability generating function of the sum of independent nonnegative integer-valued random variables is the product of their respective probability generating functions. If the chain starts at x, then

$$T_0 = T_{x-1} + (T_{x-2} - T_{x-1}) + \cdots + (T_0 - T_1)$$

is the sum of x independent random variables each having probability generating function $G(t)$. Thus the probability generating function of T_0 is $(G(t))^x$; that is,

$$(51) \qquad (G(t))^x = \sum_{n=1}^{\infty} t^n P_x(T_0 = n).$$

We will now show that

$$(52) \qquad G(t) = t\Phi(G(t)), \qquad 0 \leq t \leq 1,$$

where Φ denotes the probability generating function of f. To verify (52) we rewrite (50) as

$$G(t) = \sum_{n=0}^{\infty} t^{n+1} P_1(T_0 = n + 1) = tP(1, 0) + t \sum_{n=1}^{\infty} t^n P_1(T_0 = n + 1).$$

---

* This material is optional and can be omitted with no loss of continuity.

2.6. Queuing chain

By using successively (29) of Chapter 1, (51) of this chapter, and the formula $P(1, y) = f(y)$, $y \geq 0$, we find that

$$G(t) = tP(1, 0) + t \sum_{n=1}^{\infty} t^n \sum_{y \neq 0} P(1, y) P_y(T_0 = n)$$

$$= tP(1, 0) + t \sum_{y \neq 0} P(1, y) \sum_{n=1}^{\infty} t^n P_y(T_0 = n)$$

$$= tP(1, 0) + t \sum_{y \neq 0} P(1, y)(G(t))^y$$

$$= t \left[f(0) + \sum_{y \neq 0} f(y)(G(t))^y \right]$$

$$= t\Phi(G(t)).$$

For $0 \leq t < 1$ we can differentiate both sides of (52) and obtain

$$G'(t) = \Phi(G(t)) + tG'(t)\Phi'(G(t)).$$

Solving for $G'(t)$ we find that

(53) $$G'(t) = \frac{\Phi(G(t))}{1 - t\Phi'(G(t))}, \quad 0 \leq t < 1.$$

Now $G(t) \to 1$ and $\Phi(t) \to 1$ as $t \to 1$ and

$$\lim_{t \to 1} \Phi'(t) = \lim_{t \to 1} \sum_{x=1}^{\infty} x f(x) t^{x-1}$$

$$= \sum_{x=1}^{\infty} x f(x) = \mu.$$

By letting $t \to 1$ in (53) we see that

(54) $$\lim_{t \to 1} G'(t) = \frac{1}{1 - \mu}.$$

By definition

$$G(t) = \sum_{n=1}^{\infty} P_1(T_0 = n) t^n.$$

But since $P(1, x) = P(0, x)$, $x \geq 0$, it follows from (29) of Chapter 1 that the distribution of T_0 for a queuing chain starting in state 1 is the same as that for a chain starting in state 0. Consequently,

$$G(t) = \sum_{n=1}^{\infty} P_0(T_0 = n) t^n,$$

and hence

$$\lim_{t \to 1} G'(t) = \lim_{t \to 1} \sum_{n=1}^{\infty} n P_0(T_0 = n) t^{n-1}$$

$$= \sum_{n=1}^{\infty} n P_0(T_0 = n)$$

$$= E_0(T_0) = m_0.$$

It now follows from (54) that (49) holds. ∎

2.7. Convergence to the stationary distribution

We have seen earlier in this chapter that if X_n, $n \geq 0$, is an irreducible positive recurrent Markov chain having π as its stationary distribution, then

$$\lim_{n \to \infty} \frac{1}{n} \sum_{m=1}^{n} P^m(x, y) = \lim_{n \to \infty} \frac{G_n(x, y)}{n} = \pi(y), \qquad x, y \in \mathcal{S}.$$

In this section we will see when the stronger result

$$\lim_{n \to \infty} P^n(x, y) = \pi(y), \qquad x, y \in \mathcal{S},$$

holds and what happens when it fails to hold.

The positive integer d is said to be a *divisor* of the positive integer n if n/d is an integer. If I is a nonempty set of positive integers, the *greatest common divisor of I*, denoted by g.c.d. I, is defined to be the largest integer d such that d is a divisor of every integer in I. It follows immediately that

$$1 \leq \text{g.c.d. } I \leq \min(n : n \in I).$$

In particular, if $1 \in I$, then g.c.d. $I = 1$. The greatest common divisor of the set of even positive integers is 2.

Let x be a state of a Markov chain such that $P^n(x, x) > 0$ for some $n \geq 1$, i.e., such that $\rho_{xx} = P_x(T_x < \infty) > 0$. We define its *period* d_x by

$$d_x = \text{g.c.d. } \{n \geq 1 : P^n(x, x) > 0\}.$$

Then

$$1 \leq d_x \leq \min(n \geq 1 : P^n(x, x) > 0).$$

If $P(x, x) > 0$, then $d_x = 1$.

If x and y are two states, each of which leads to the other, then $d_x = d_y$. For let n_1 and n_2 be positive integers such that

$$P^{n_1}(x, y) > 0 \quad \text{and} \quad P^{n_2}(y, x) > 0.$$

2.7. Convergence to the stationary distribution

Then
$$P^{n_1+n_2}(x, x) \geq P^{n_1}(x, y)P^{n_2}(y, x) > 0,$$
and hence d_x is a divisor of $n_1 + n_2$. If $P^n(y, y) > 0$, then
$$P^{n_1+n+n_2}(x, x) \geq P^{n_1}(x, y)P^n(y, y)P^{n_2}(y, x) > 0,$$
so that d_x is a divisor of $n_1 + n + n_2$. Since d_x is a divisor of $n_1 + n_2$, it must be a divisor of n. Thus d_x is a divisor of all numbers in the set $\{n \geq 1: P^n(y, y) > 0\}$. Since d_y is the largest such divisor, we conclude that $d_x \leq d_y$. Similarly $d_y \leq d_x$, and hence $d_x = d_y$.

We have shown, in other words, that the states in an irreducible Markov chain have common period d. We say that the chain is *periodic with period* d if $d > 1$ and *aperiodic* if $d = 1$. A simple sufficient condition for an irreducible Markov chain to be aperiodic is that $P(x, x) > 0$ for some $x \in \mathscr{S}$. Since $P(0, 0) = f(0) > 0$ for an irreducible queuing chain, such a chain is necessarily aperiodic.

Example 7. Determine the period of an irreducible birth and death chain.

If some $r_x > 0$, then $P(x, x) = r_x > 0$, and the birth and death chain is aperiodic. In particular, the modified Ehrenfest chain in Example 3 is aperiodic.

Suppose $r_x = 0$ for all x. Then in one transition the state of the chain changes either from an odd numbered state to an even numbered state or from an even numbered state to an odd numbered state. In particular, a chain can return to its initial state only after an even number of transitions. Thus the period of the chain is 2 or a multiple of 2. Since
$$P^2(0, 0) = p_0 q_1 > 0,$$
we conclude that the chain is periodic with period 2. In particular, the Ehrenfest chain introduced in Example 2 of Chapter 1 is periodic with period 2.

Theorem 7 *Let X_n, $n \geq 0$, be an irreducible positive recurrent Markov chain having stationary distribution π. If the chain is aperiodic,*

(55) $$\lim_{n \to \infty} P^n(x, y) = \pi(y), \quad x, y \in \mathscr{S}.$$

If the chain is periodic with period d, then for each pair x, y of states in $\mathscr{S}$ there is an integer r, $0 \leq r < d$, such that $P^n(x, y) = 0$ unless $n = md + r$ for some nonnegative integer m, and

(56) $$\lim_{m \to \infty} P^{md+r}(x, y) = d\pi(y).$$

For an illustration of the second half of this theorem, consider an irreducible positive recurrent birth and death chain which is periodic with period 2. If $y - x$ is even, then $P^{2m+1}(x, y) = 0$ for all $m \geq 0$ and

$$\lim_{m \to \infty} P^{2m}(x, y) = 2\pi(y).$$

If $y - x$ is odd, then $P^{2m}(x, y) = 0$ for all $m \geq 1$ and

$$\lim_{m \to \infty} P^{2m+1}(x, y) = 2\pi(y).$$

We will prove this theorem in an appendix to this chapter, which can be omitted with no loss of continuity.

Example 8. Determine the asymptotic behavior of the matrix P^n for the transition matrix P
(a) from Example 3,
(b) from Example 2.

(a) The transition matrix P from Example 3 corresponds to an aperiodic irreducible Markov chain on $\{0, 1, 2, 3\}$ having the stationary distribution given by

$$\pi(0) = \tfrac{1}{8}, \qquad \pi(1) = \tfrac{3}{8}, \qquad \pi(2) = \tfrac{3}{8}, \qquad \text{and} \qquad \pi(3) = \tfrac{1}{8}.$$

It follows from Theorem 7 that for n large

$$P^n \doteq \begin{bmatrix} \tfrac{1}{8} & \tfrac{3}{8} & \tfrac{3}{8} & \tfrac{1}{8} \\ \tfrac{1}{8} & \tfrac{3}{8} & \tfrac{3}{8} & \tfrac{1}{8} \\ \tfrac{1}{8} & \tfrac{3}{8} & \tfrac{3}{8} & \tfrac{1}{8} \\ \tfrac{1}{8} & \tfrac{3}{8} & \tfrac{3}{8} & \tfrac{1}{8} \end{bmatrix}.$$

(b) The transition matrix P from Example 2 corresponds to a periodic irreducible Markov chain on $\{0, 1, 2, 3\}$ having period 2 and the same stationary distribution as the chain in Example 3. From the discussion following the statement of Theorem 7, we conclude that for n large and even

$$P^n \doteq \begin{bmatrix} \tfrac{1}{4} & 0 & \tfrac{3}{4} & 0 \\ 0 & \tfrac{3}{4} & 0 & \tfrac{1}{4} \\ \tfrac{1}{4} & 0 & \tfrac{3}{4} & 0 \\ 0 & \tfrac{3}{4} & 0 & \tfrac{1}{4} \end{bmatrix},$$

while for n large and odd

$$P^n \doteq \begin{bmatrix} 0 & \tfrac{3}{4} & 0 & \tfrac{1}{4} \\ \tfrac{1}{4} & 0 & \tfrac{3}{4} & 0 \\ 0 & \tfrac{3}{4} & 0 & \tfrac{1}{4} \\ \tfrac{1}{4} & 0 & \tfrac{3}{4} & 0 \end{bmatrix}.$$

APPENDIX

2.8. Proof of convergence

We will first prove Theorem 7 in the aperiodic case. Consider an aperiodic, irreducible, positive recurrent Markov chain having transition function P, state space $\mathscr{S}$, and stationary distribution π. We will now verify that the conclusion of Theorem 7 holds for such a chain.

Choose $a \in \mathscr{S}$ and let I be the set of positive integers defined by

$$I = \{n > 0 : P^n(a, a) > 0\}.$$

Then

(i) g.c.d. $I = 1$;
(ii) if $m \in I$ and $n \in I$, then $m + n \in I$.

Property (ii) follows from the inequality

$$P^{m+n}(a, a) \geq P^m(a, a)P^n(a, a).$$

Properties (i) and (ii) imply that there is a positive integer n_1 such that $n \in I$ for all $n \geq n_1$. For completeness we will prove this number theoretic result in Section 2.8.2. Using this result we conclude that $P^n(a, a) > 0$ for $n \geq n_1$.

Let x and y be any pair of states in $\mathscr{S}$. Since the chain is irreducible, there exist positive integers n_2 and n_3 such that

$$P^{n_2}(x, a) > 0 \quad \text{and} \quad P^{n_3}(a, y) > 0.$$

Then for $n \geq n_1$

$$P^{n_2+n+n_3}(x, y) \geq P^{n_2}(x, a)P^n(a, a)P^{n_3}(a, y) > 0.$$

We have shown, in other words, that for every pair x, y of states in $\mathscr{S}$ there is a positive integer n_0 such that

(57) $$P^n(x, y) > 0, \quad n \geq n_0.$$

Set

$$\mathscr{S}^2 = \{(x, y) : x \in \mathscr{S} \text{ and } y \in \mathscr{S}\}.$$

Then $\mathscr{S}^2$ is the set of ordered pairs of elements in $\mathscr{S}$. We will consider a Markov chain (X_n, Y_n) having state space $\mathscr{S}^2$ and transition function P_2 defined by

$$P_2((x_0, y_0), (x, y)) = P(x_0, x)P(y_0, y).$$

It follows that $X_n, n \geq 0$, and $Y_n, n \geq 0$, are each Markov chains having transition function P, and the successive transitions of the X_n chain and the Y_n chain are chosen independently of each other.

We will now develop properties of the Markov chain (X_n, Y_n). In particular, we will show that this chain is an aperiodic, irreducible, positive recurrent Markov chain. We will then use this chain to verify the conclusion of the theorem.

Choose $(x_0, y_0) \in \mathscr{S}^2$ and $(x, y) \in \mathscr{S}^2$. By (57) there is an $n_0 > 0$ such that
$$P^n(x_0, x) > 0 \quad \text{and} \quad P^n(y_0, y) > 0, \quad n \geq n_0.$$

Then

(58) $\quad P_2^n((x_0, y_0), (x, y)) = P^n(x_0, x) P^n(y_0, y) > 0, \quad n \geq n_0.$

We conclude from (58) that the chain is both irreducible and aperiodic.

The distribution π_2 on $\mathscr{S}^2$ defined by $\pi_2(x_0, y_0) = \pi(x_0)\pi(y_0)$ is a stationary distribution. For

$$\sum_{(x_0, y_0) \in \mathscr{S}^2} \pi_2(x_0, y_0) P_2((x_0, y_0), (x, y))$$
$$= \sum_{x_0 \in \mathscr{S}} \sum_{y_0 \in \mathscr{S}} \pi(x_0)\pi(y_0) P(x_0, x) P(y_0, y)$$
$$= \left(\sum_{x_0 \in \mathscr{S}} \pi(x_0) P(x_0, x) \right) \left(\sum_{y_0 \in \mathscr{S}} \pi(y_0) P(y_0, y) \right)$$
$$= \pi(x)\pi(y) = \pi_2(x, y).$$

Thus the chain on $\mathscr{S}^2$ is positive recurrent; in particular, it is recurrent.

Set
$$T = \min(n > 0 : X_n = Y_n).$$

Choose $a \in \mathscr{S}$. Since the (X_n, Y_n) chain is recurrent,
$$T_{(a,a)} = \min(n > 0 : (X_n, Y_n) = (a, a))$$
is finite with probability one. Clearly $T \leq T_{(a,a)}$, and hence T is finite with probability one.

For any $n \geq 1$ (regardless of the distribution of (X_0, Y_0))

(59) $\quad P(X_n = y, T \leq n) = P(Y_n = y, T \leq n), \quad y \in \mathscr{S}.$

This formula is intuitively reasonable since the two chains are indistinguishable for $n \geq T$. To make this argument precise, we choose $1 \leq m \leq n$. Then for $z \in \mathscr{S}$

(60) $\quad P(X_n = y \mid T = m, X_m = Y_m = z)$
$$= P(Y_n = y \mid T = m, X_m = Y_m = z),$$

since both conditional probabilities equal $P^{n-m}(z, y)$. Now the event $\{T = n\}$ is the union of the disjoint events
$$\{T = m, X_m = Y_m = z\}, \quad 1 \leq m \leq n \quad \text{and} \quad z \in \mathscr{S},$$

2.8. Proof of convergence

so it follows from (60) and Exercise 4(d) of Chapter 1 that

$$P(X_n = y \mid T \le n) = P(Y_n = y \mid T \le n)$$

and hence that (59) holds.

Equation (59) implies that

$$\begin{aligned}P(X_n = y) &= P(X_n = y, T \le n) + P(X_n = y, T > n) \\ &= P(Y_n = y, T \le n) + P(X_n = y, T > n) \\ &\le P(Y_n = y) + P(T > n)\end{aligned}$$

and similarly that

$$P(Y_n = y) \le P(X_n = y) + P(T > n).$$

Therefore for $n \ge 1$

(61) $\qquad |P(X_n = y) - P(Y_n = y)| \le P(T > n), \qquad y \in \mathscr{S}.$

Since T is finite with probability one,

(62) $\qquad \lim_{n \to \infty} P(T > n) = 0.$

We conclude from (61) and (62) that

(63) $\qquad \lim_{n \to \infty} (P(X_n = y) - P(Y_n = y)) = 0, \qquad y \in \mathscr{S}.$

Using (63), we can easily complete the proof of Theorem 7. Choose $x \in \mathscr{S}$ and let the initial distribution of (X_n, Y_n) be such that $P(X_0 = x) = 1$ and

$$P(Y_0 = y_0) = \pi(y_0), \qquad y_0 \in \mathscr{S}.$$

Since $X_n, n \ge 0,$ and $Y_n, n \ge 0,$ are each Markov chains with transition function P, we see that

(64) $\qquad P(X_n = y) = P^n(x, y), \qquad y \in \mathscr{S},$

and

(65) $\qquad P(Y_n = y) = \pi(y), \qquad y \in \mathscr{S}.$

Thus by (63)–(65)

$$\lim_{n \to \infty} (P^n(x, y) - \pi(y)) = \lim_{n \to \infty} (P(X_n = y) - P(Y_n = y)) = 0,$$

and hence the conclusion of Theorem 7 holds.

2.8.1. Periodic case.
We first consider a slight extension of Theorem 7 in the aperiodic case. Let C be an irreducible closed set of positive recurrent states such that each state in C has period 1, and let π

be the unique stationary distribution concentrated on C. By looking at the Markov chain restricted to C, we conclude that

$$\lim_{n \to \infty} P^n(x, y) = \pi(y) = \frac{1}{m_y}, \quad x, y \in C.$$

In particular, if y is any positive recurrent state having period 1, then by letting C be the irreducible closed set containing y, we see that

(66) $$\lim_{n \to \infty} P^n(y, y) = \frac{1}{m_y}.$$

We now proceed with the proof of Theorem 7 in the periodic case. Let X_n, $n \geq 0$, be an irreducible positive recurrent Markov chain which is periodic with period $d > 1$. Set $Y_m = X_{md}$, $m \geq 0$. Then Y_m, $m \geq 0$, is a Markov chain having transition function $Q = P^d$. Choose $y \in \mathscr{S}$. Then

$$\text{g.c.d. } \{m \mid Q^m(y, y) > 0\} = \text{g.c.d. } \{m \mid P^{md}(y, y) > 0\}$$

$$= \frac{1}{d} \text{g.c.d. } \{n \mid P^n(y, y) > 0\}$$

$$= 1.$$

Thus all states have period 1 with respect to the Y_m chain.

Let the X_n chain and hence also the Y_m chain start at y. Since the X_n chain first returns to y at some multiple of d, it follows that the expected return time to y for the Y_m chain is $d^{-1}m_y$, where m_y is the expected return time to y for the X_n chain. In particular, y is a positive recurrent state for a Markov chain having transition function Q. By applying (66) to this transition function we conclude that

$$\lim_{m \to \infty} Q^m(y, y) = \frac{d}{m_y} = d\pi(y),$$

and thus that

(67) $$\lim_{m \to \infty} P^{md}(y, y) = d\pi(y), \quad y \in \mathscr{S}.$$

Let x and y be any pair of states in $\mathscr{S}$ and set

$$r_1 = \min (n : P^n(x, y) > 0).$$

Then, in particular, $P^{r_1}(x, y) > 0$. We will show that $P^n(x, y) > 0$ only if $n - r_1$ is an integral multiple of d.

Choose n_1 such that $P^{n_1}(y, x) > 0$. Then

$$P^{r_1 + n_1}(y, y) \geq P^{n_1}(y, x) P^{r_1}(x, y) > 0,$$

2.8. Proof of convergence

and hence $r_1 + n_1$ is an integral multiple of d. If $P^n(x, y) > 0$, then by the same argument $n + n_1$ is an integral multiple of d, and therefore so is $n - r_1$. Thus, $n = kd + r_1$ for some nonnegative integer k.

There is a nonnegative integer m_1 such that $r_1 = m_1 d + r$, where $0 \le r < d$. We conclude that

(68) $\qquad P^n(x, y) = 0 \qquad \text{unless} \qquad n = md + r$

for some nonnegative integer m. It follows from (68) and from (28) of Chapter 1 that

(69) $\qquad P^{md+r}(x, y) = \sum_{k=0}^{m} P_x(T_y = kd + r) P^{(m-k)d}(y, y).$

Set

$$a_m(k) = \begin{cases} P^{(m-k)d}(y, y), & 0 \le k \le m, \\ 0, & k > m. \end{cases}$$

Then by (67) for each fixed k

$$\lim_{m \to \infty} a_m(k) = d\pi(y).$$

We can apply the bounded convergence theorem (with $\mathscr{S}$ replaced by $\{0, 1, 2, \ldots\}$) to conclude from (69) that

$$\lim_{m \to \infty} P^{md+r}(x, y) = d\pi(y) \sum_{k=0}^{\infty} P_x(T_y = kd + r)$$

$$= d\pi(y) P_x(T_y < \infty)$$

$$= d\pi(y),$$

and hence that (56) holds. This completes the proof of Theorem 7. ∎

2.8.2. A result from number theory.
Let I be a nonempty set of positive integers such that
   (i) g.c.d. $I = 1$;
   (ii) if m and n are in I, then $m + n$ is in I.
Then there is an n_0 such that $n \in I$ for all $n \ge n_0$.

We will first prove that I contains two consecutive integers. Suppose otherwise. Then there is an integer $k \ge 2$ and an $n_1 \in I$ such that $n_1 + k \in I$ and any two distinct integers in I differ by at least k. It follows from property (i) that there is an $n \in I$ such that k is not a divisor of n. We can write

$$n = mk + r,$$

where m is a nonnegative integer and $0 < r < k$. It follows from property (ii) that $(m + 1)(n_1 + k)$ and $n + (m + 1)n_1$ are each in I. Their difference is

$$(m + 1)(n_1 + k) - n - (m + 1)n_1 = k + mk - n = k - r,$$

which is positive and smaller than k. This contradicts the definition of k.

We have shown that I contains two consecutive integers, say n_1 and $n_1 + 1$. Let $n \geq n_1^2$. Then there are nonnegative integers m and r such that $0 \leq r < n_1$ and

$$n - n_1^2 = mn_1 + r.$$

Thus

$$n = r(n_1 + 1) + (n_1 - r + m)n_1,$$

which is in I by property (ii). This shows that $n \in I$ for all

$$n \geq n_0 = n_1^2. \qquad \blacksquare$$

Exercises

1 Consider a Markov chain having state space $\{0, 1, 2\}$ and transition matrix

$$\begin{array}{c} \\ 0 \\ 1 \\ 2 \end{array} \begin{array}{ccc} 0 & 1 & 2 \\ \left[\begin{array}{ccc} .4 & .4 & .2 \\ .3 & .4 & .3 \\ .2 & .4 & .4 \end{array} \right]. \end{array}$$

Show that this chain has a unique stationary distribution π and find π.

2 Consider a Markov chain having transition function P such that $P(x, y) = \alpha_y$, $x \in \mathscr{S}$ and $y \in \mathscr{S}$, where the α_y's are constants. Show that the chain has a unique stationary distribution π, given by $\pi(y) = \alpha_y$, $y \in \mathscr{S}$.

3 Let π be a stationary distribution of a Markov chain. Show that if $\pi(x) > 0$ and x leads to y, then $\pi(y) > 0$.

4 Let π be a stationary distribution of a Markov chain. Suppose that y and z are two states such that for some constant c

$$P(x, y) = cP(x, z), \qquad x \in \mathscr{S}.$$

Show that $\pi(y) = c\pi(z)$.

5 Let π_0 and π_1 be distinct stationary distributions for a Markov chain.
(a) Show that for $0 \leq \alpha \leq 1$, the function π_α defined by

$$\pi_\alpha(x) = (1 - \alpha)\pi_0(x) + \alpha\pi_1(x), \qquad x \in \mathscr{S},$$

is a stationary distribution.

(b) Show that distinct values of α determine distinct stationary distributions π_α. *Hint:* Choose $x_0 \in \mathscr{S}$ such that $\pi_0(x_0) \neq \pi_1(x_0)$ and show that $\pi_\alpha(x_0) = \pi_\beta(x_0)$ implies that $\alpha = \beta$.

6 Consider a birth and death chain on the nonnegative integers and suppose that $p_0 = 1, p_x = p > 0$ for $x \geq 1$, and $q_x = q = 1 - p > 0$ for $x \geq 1$. Find the stationary distribution when it exists.

7 (a) Find the stationary distribution of the Ehrenfest chain.
   (b) Find the mean and variance of this distribution.

8 For general d, find the transition function of the modified Ehrenfest chain introduced in Example 3, and show that this chain has the same stationary distribution as does the original Ehrenfest chain.

9 Find the stationary distribution of the birth and death chain described in Exercise 2 of Chapter 1. *Hint:* Use the formula

$$\binom{d}{0}^2 + \cdots + \binom{d}{d}^2 = \binom{2d}{d}.$$

10 Let X_n, $n \geq 0$, be a positive recurrent irreducible birth and death chain, and suppose that X_0 has the stationary distribution π. Show that

$$P(X_0 = y \mid X_1 = x) = P(x, y), \qquad x, y \in \mathscr{S}.$$

Hint: Use the definition of π_x given by (9).

11 Let X_n, $n \geq 0$, be the Markov chain introduced in Section 2.2.2. Show that if X_0 has a Poisson distribution with parameter t, then X_n has a Poisson distribution with parameter

$$tp^n + \frac{\lambda}{q}(1 - p^n).$$

12 Let X_n, $n \geq 0$, be as in Exercise 11. Show that

$$E_x(X_n) = xp^n + \frac{\lambda}{q}(1 - p^n).$$

Hint: Use the result of Exercise 11 and equate coefficients of t^x in the appropriate power series.

13 Let $X_n, n \geq 0$, be as in Exercise 11 and suppose that X_0 has the stationary distribution. Use the result of Exercise 12 to find $\text{cov}(X_m, X_{m+n})$, $m \geq 0$ and $n \geq 0$.

14 Consider a Markov chain on the nonnegative integers having transition function P given by $P(x, x + 1) = p$ and $P(x, 0) = 1 - p$, where $0 < p < 1$. Show that this chain has a unique stationary distribution π and find π.

15 The transition function of a Markov chain is called *doubly stochastic* if

$$\sum_{x \in \mathcal{S}} P(x, y) = 1, \qquad y \in \mathcal{S}.$$

What is the stationary distribution of an irreducible Markov chain having $d < \infty$ states and a doubly stochastic transition function?

16 Consider an irreducible Markov chain having finite state space $\mathcal{S}$, transition function P such that $P(x, x) = 0$, $x \in \mathcal{S}$ and stationary distribution π. Let p_x, $x \in \mathcal{S}$, be such that $0 < p_x < 1$, and let $Q(x, y)$, $x \in \mathcal{S}$ and $y \in \mathcal{S}$, be defined by

$$Q(x, x) = 1 - p_x$$

and

$$Q(x, y) = p_x P(x, y), \qquad y \neq x.$$

Show that Q is the transition function of an irreducible Markov chain having state space $\mathcal{S}$ and stationary distribution π', defined by

$$\pi'(x) = \frac{p_x^{-1} \pi(x)}{\sum_{y \in \mathcal{S}} p_y^{-1} \pi(y)}, \qquad x \in \mathcal{S}.$$

The interpretation of the chain with transition function Q is that starting from x, it has probability $1 - p_x$ of remaining in x and probability p_x of jumping according to the transition function P.

17 Consider the Ehrenfest chain. Suppose that initially all of the balls are in the second box. Find the expected amount of time until the system returns to that state. *Hint:* Use the result of Exercise 7(a).

18 A particle moves according to a Markov chain on $\{1, 2, \ldots, c + d\}$, where c and d are positive integers. Starting from any one of the first c states, the particle jumps in one transition to a state chosen uniformly from the last d states; starting from any of the last d states, the particle jumps in one transition to a state chosen uniformly from the first c states.

(a) Show that the chain is irreducible.
(b) Find the stationary distribution.

19 Consider a Markov chain having the transition matrix given by Exercise 19 of Chapter 1.

(a) Find the stationary distribution concentrated on each of the irreducible closed sets.
(b) Find $\lim_{n \to \infty} G_n(x, y)/n$.

20 Consider a Markov chain having transition matrix as in Exercise 20 of Chapter 1.

(a) Find the stationary distribution concentrated on each of the irreducible closed sets.
(b) Find $\lim_{n \to \infty} G_n(x, y)/n$.

Exercises

21 Let X_n, $n \geq 0$, be the Ehrenfest chain with $d = 4$ and $X_0 = 0$.
   (a) Find the approximate distribution of X_n for n large and even.
   (b) Find the approximate distribution of X_n for n large and odd.

22 Consider a Markov chain on $\{0, 1, 2\}$ having transition matrix

$$P = \begin{array}{c} 0 \\ 1 \\ 2 \end{array} \begin{bmatrix} 0 & 1 & 2 \\ 0 & 0 & 1 \\ 1 & 0 & 0 \\ \frac{1}{2} & \frac{1}{2} & 0 \end{bmatrix}.$$

   (a) Show that the chain is irreducible.
   (b) Find the period.
   (c) Find the stationary distribution.

23 Consider a Markov chain on $\{0, 1, 2, 3, 4\}$ having transition matrix

$$P = \begin{array}{c} 0 \\ 1 \\ 2 \\ 3 \\ 4 \end{array} \begin{bmatrix} 0 & \frac{1}{3} & \frac{2}{3} & 0 & 0 \\ 0 & 0 & 0 & \frac{1}{4} & \frac{3}{4} \\ 0 & 0 & 0 & \frac{1}{4} & \frac{3}{4} \\ 1 & 0 & 0 & 0 & 0 \\ 1 & 0 & 0 & 0 & 0 \end{bmatrix}.$$

   (a) Show that the chain is irreducible.
   (b) Find the period.
   (c) Find the stationary distribution.

3 Markov Pure Jump Processes

Consider again a system that at any time can be in one of a finite or countably infinite set $\mathscr{S}$ of states. We call $\mathscr{S}$ the *state space* of the system. In Chapters 1 and 2 we studied the behavior of such systems at integer times. In this chapter we will study the behavior of such systems over all times $t \geq 0$.

3.1. Construction of jump processes

Consider a system starting in state x_0 at time 0. We suppose that the system remains in state x_0 until some positive time τ_1, at which time the system jumps to a new state $x_1 \neq x_0$. We allow the possibility that the system remains permanently in state x_0, in which case we set $\tau_1 = \infty$. If τ_1 is finite, upon reaching x_1 the system remains there until some time $\tau_2 > \tau_1$ when it jumps to state $x_2 \neq x_1$. If the system never leaves x_1, we set $\tau_2 = \infty$. This procedure is repeated indefinitely. If some $\tau_m = \infty$, we set $\tau_n = \infty$ for $n > m$.

Let $X(t)$ denote the state of the system at time t, defined by

(1) $$X(t) = \begin{cases} x_0, & 0 \leq t < \tau_1, \\ x_1, & \tau_1 \leq t < \tau_2, \\ x_2, & \tau_2 \leq t < \tau_3, \\ \vdots & \end{cases}$$

The process defined by (1) is called a *jump process*. At first glance it might appear that (1) defines $X(t)$ for all $t \geq 0$. But this is not necessarily the case.

Consider, for example, a ball bouncing on the floor. Let the state of the system be the number of bounces it has made. We make the physically reasonable assumption that the time in seconds between the nth bounce and the $(n+1)$th bounce is 2^{-n}. Then $x_n = n$ and

$$\tau_n = 1 + \frac{1}{2} + \cdots + \frac{1}{2^{n-1}} = 2 - \frac{1}{2^{n-1}}.$$

3.1. Construction of jump processes

We see that $\tau_n < 2$ and $\tau_n \to 2$ as $n \to \infty$. Thus (1) defines $X(t)$ only for $0 \le t < 2$. By the time $t = 2$ the ball will have made an infinite number of bounces. In this case it would be appropriate to define $X(t) = \infty$ for $t \ge 2$.

In general, if

(2) $$\lim_{n \to \infty} \tau_n < \infty,$$

we say that the $X(t)$ process *explodes*. If the $X(t)$ process does not explode, i.e., if

(3) $$\lim_{n \to \infty} \tau_n = \infty,$$

then (1) *does* define $X(t)$ for all $t \ge 0$.

We will now specify a probability structure for such a jump process. We suppose that all states are of one of two types, *absorbing* or *non-absorbing*. Once the process reaches an absorbing state, it remains there permanently. With each non-absorbing state x, there is associated a distribution function $F_x(t)$, $-\infty < t < \infty$, which vanishes for $t \le 0$, and transition probabilities Q_{xy}, $y \in \mathcal{S}$, which are nonnegative and such that $Q_{xx} = 0$ and

(4) $$\sum_y Q_{xy} = 1.$$

A process starting at x remains there for a random length of time τ_1 having distribution function F_x and then jumps to state $X(\tau_1) = y$ with probability Q_{xy}, $y \in \mathcal{S}$. We assume that τ_1 and $X(\tau_1)$ are chosen independently of each other, i.e., that

$$P_x(\tau_1 \le t, X(\tau_1) = y) = F_x(t) Q_{xy}.$$

Here, as in the previous chapters, we use the notation $P_x(\ )$ and $E_x(\ )$ to denote probabilities of events and expectations of random variables defined in terms of a process initially in state x. Whenever and however the process jumps to a state y, it acts just as a process starting initially at y. For example, if x and y are both non-absorbing states,

$$P_x(\tau_1 \le s, X(\tau_1) = y, \tau_2 - \tau_1 \le t, X(\tau_2) = z) = F_x(s) Q_{xy} F_y(t) Q_{yz}.$$

Similar formulas hold for events defined in terms of three or more jumps. If x is an absorbing state, we set $Q_{xy} = \delta_{xy}$, where

$$\delta_{xy} = \begin{cases} 1, & y = x, \\ 0, & y \ne x. \end{cases}$$

Equation (4) now holds for all $x \in \mathcal{S}$.

We say that the jump process is *pure* or *non-explosive* if (3) holds with probability one regardless of the starting point. Otherwise we say the process is *explosive*. If the state space $\mathscr{S}$ is finite, the jump process is necessarily non-explosive. It is easy to construct examples having an infinite state space which are explosive. Such processes, however, are unlikely to arise in practical applications. At any rate, to keep matters simple we assume that our process is non-explosive. The set of probability zero where (3) fails to hold can safely be ignored. We see from (1) that $X(t)$ is then defined for all $t \geq 0$.

Let $P_{xy}(t)$ denote the probability that a process starting in state x will be in state y at time t. Then

$$P_{xy}(t) = P_x(X(t) = y)$$

and

$$\sum_y P_{xy}(t) = 1.$$

In particular, $P_{xy}(0) = \delta_{xy}$. We can also choose the initial state x according to an *initial distribution* $\pi_0(x)$, $x \in \mathscr{S}$, where $\pi_0(x) \geq 0$ and

$$\sum_x \pi_0(x) = 1.$$

In this case,

$$P(X(t) = y) = \sum_x \pi_0(x) P_{xy}(t).$$

The *transition function* $P_{xy}(t)$ cannot be used directly to obtain such probabilities as

$$P(X(t_1) = x_1, \ldots, X(t_n) = x_n)$$

unless the jump process satisfies the *Markov property*, which states that for $0 \leq s_1 \leq \cdots \leq s_n \leq s \leq t$ and $x_1, \ldots, x_n, x, y \in \mathscr{S}$,

$$P(X(t) = y \mid X(s_1) = x_1, \ldots, X(s_n) = x_n, X(s) = x) = P_{xy}(t - s).$$

By a *Markov pure jump process* we mean a pure jump process that satisfies the Markov property. It can be shown, although not at the level of this book, that a pure jump process is Markovian if and only if all non-absorbing states x are such that

$$P_x(\tau_1 > t + s \mid \tau_1 > s) = P_x(\tau_1 > t), \qquad s, t \geq 0,$$

i.e., such that

(5) $$\frac{1 - F_x(t + s)}{1 - F_x(s)} = 1 - F_x(t), \qquad s, t \geq 0.$$

Now a distribution function F_x satisfies (5) if and only if it is an exponential distribution function (see Chapter 5 of *Introduction to Probability Theory*). We conclude that a pure jump process is Markovian if and only if F_x is an exponential distribution for all non-absorbing states x.

3.1. Construction of jump processes

Let $X(t)$, $0 \leq t < \infty$, be a Markov pure jump process. If x is a non-absorbing state, then F_x has an exponential density f_x. Let q_x denote the parameter of this density. Then $q_x = 1/E_x(\tau_1) > 0$ and

$$f_x(t) = \begin{cases} q_x e^{-q_x t}, & t \geq 0, \\ 0, & t < 0. \end{cases}$$

Observe that

$$P_x(\tau_1 \geq t) = \int_t^\infty q_x e^{-q_x s}\, ds = e^{-q_x t}, \qquad t \geq 0.$$

If x is an absorbing state, we set $q_x = 0$.

It follows from the Markov property that for $0 \leq t_1 \leq \cdots \leq t_n$ and $x_1, \ldots, x_n$ in $\mathscr{S}$,

(6) $\quad P(X(t_1) = x_1, \ldots, X(t_n) = x_n)$
$$= P(X(t_1) = x_1) P_{x_1 x_2}(t_2 - t_1) \cdots P_{x_{n-1} x_n}(t_n - t_{n-1}).$$

In particular, for $s \geq 0$ and $t \geq 0$

$$P_x(X(t) = z, X(t + s) = y) = P_{xz}(t) P_{zy}(s).$$

Since

$$P_{xy}(t + s) = \sum_z P_x(X(t) = z, X(t + s) = y),$$

we conclude that

(7) $\quad P_{xy}(t + s) = \sum_z P_{xz}(t) P_{zy}(s), \qquad s \geq 0 \text{ and } t \geq 0.$

Equation (7) is known as the *Chapman–Kolmogorov* equation.

The transition function $P_{xy}(t)$ satisfies the integral equation

(8) $\quad P_{xy}(t) = \delta_{xy} e^{-q_x t} + \int_0^t q_x e^{-q_x s} \left(\sum_{z \neq x} Q_{xz} P_{zy}(t - s) \right) ds, \qquad t \geq 0,$

which we will now verify. If x is an absorbing state, (8) reduces to the obvious fact that

$$P_{xy}(t) = \delta_{xy}, \qquad t \geq 0.$$

Suppose x is not an absorbing state. Then for a process starting at x, the event $\{\tau_1 \leq t, X(\tau_1) = z \text{ and } X(t) = y\}$ occurs if and only if the first jump occurs at some time $s \leq t$ and takes the process to z, and the process goes from z to y in the remaining $t - s$ units of time. Thus

$$P_x(\tau_1 \leq t, X(\tau_1) = z \text{ and } X(t) = y) = \int_0^t q_x e^{-q_x s} Q_{xz} P_{zy}(t - s)\, ds,$$

so

$$P_x(\tau_1 \le t \text{ and } X(t) = y) = \sum_{z \ne x} P_x(\tau_1 \le t, X(\tau_1) = z \text{ and } X(t) = y)$$
$$= \int_0^t q_x e^{-q_x s} \left(\sum_{z \ne x} Q_{xz} P_{zy}(t-s) \right) ds.$$

Also

$$P_x(\tau_1 > t \text{ and } X(t) = y) = \delta_{xy} P_x(\tau_1 > t)$$
$$= \delta_{xy} e^{-q_x t}.$$

Consequently,

$$P_{xy}(t) = P_x(X(t) = y)$$
$$= P_x(\tau_1 > t \text{ and } X(t) = y) + P_x(\tau_1 \le t \text{ and } X(t) = y)$$
$$= \delta_{xy} e^{-q_x t} + \int_0^t q_x e^{-q_x s} \left(\sum_{z \ne x} Q_{xz} P_{zy}(t-s) \right) ds,$$

as claimed. Replacing s by $t - s$ in the integral in (8), we can rewrite (8) as

(9) $\quad P_{xy}(t) = \delta_{xy} e^{-q_x t} + q_x e^{-q_x t} \int_0^t e^{q_x s} \left(\sum_{z \ne x} Q_{xz} P_{zy}(s) \right) ds, \quad t \ge 0.$

It follows from (9) that $P_{xy}(t)$ is continuous in t for $t \ge 0$. Therefore the integrand in (9) is a continuous function, so we can differentiate the right side. We obtain

(10) $\quad P'_{xy}(t) = -q_x P_{xy}(t) + q_x \sum_{z \ne x} Q_{xz} P_{zy}(t), \quad t \ge 0.$

In particular,

$$P'_{xy}(0) = -q_x P_{xy}(0) + q_x \sum_{z \ne x} Q_{xz} P_{zy}(0)$$
$$= -q_x \delta_{xy} + q_x \sum_{z \ne x} Q_{xz} \delta_{zy}$$
$$= -q_x \delta_{xy} + q_x Q_{xy}.$$

Set

(11) $\quad q_{xy} = P'_{xy}(0), \quad x, y \in \mathscr{S}.$

Then

(12) $\quad q_{xy} = \begin{cases} -q_x, & y = x, \\ q_x Q_{xy}, & y \ne x. \end{cases}$

It follows from (12) that

(13) $\quad \sum_{y \ne x} q_{xy} = q_x = -q_{xx}.$

The quantities q_{xy}, $x \in \mathcal{S}$ and $y \in \mathcal{S}$, are called the *infinitesimal parameters* of the process. These parameters determine q_x and Q_{xy}, and thus by our construction determine a unique Markov pure jump process. We can rewrite (10) in terms of the infinitesimal parameters as

(14) $$P'_{xy}(t) = \sum_z q_{xz} P_{zy}(t), \qquad t \geq 0.$$

This equation is known as the *backward equation*.

If $\mathcal{S}$ is finite, we can differentiate the Chapman-Kolmogorov equation with respect to s, obtaining

(15) $$P'_{xy}(t + s) = \sum_z P_{xz}(t) P'_{zy}(s), \qquad s \geq 0 \text{ and } t \geq 0.$$

In particular,

$$P'_{xy}(t) = \sum_z P_{xz}(t) P'_{zy}(0), \qquad t \geq 0,$$

or equivalently,

(16) $$P'_{xy}(t) = \sum_z P_{xz}(t) q_{zy}, \qquad t \geq 0.$$

Formula (16) is known as the *forward equation*. It can be shown that (15) and (16) hold even if $\mathcal{S}$ is infinite, but the proofs are not easy and will be omitted.

In Section 3.2 we will describe some examples in which the backward or forward equation can be used to find explicit formulas for $P_{xy}(t)$.

3.2. Birth and death processes

Let $\mathcal{S} = \{0, 1, \ldots, d\}$ or $\mathcal{S} = \{0, 1, 2, \ldots\}$. By a *birth and death process* on $\mathcal{S}$ we mean a Markov pure jump process on $\mathcal{S}$ having infinitesimal parameters q_{xy} such that

$$q_{xy} = 0, \qquad |y - x| > 1.$$

Thus a birth and death process starting at x can in one jump go only to the states $x - 1$ or $x + 1$.

The parameters $\lambda_x = q_{x,x+1}$, $x \in \mathcal{S}$, and $\mu_x = q_{x,x-1}$, $x \in \mathcal{S}$, are called respectively the *birth rates* and *death rates* of the process. The parameters q_x and Q_{xy} of the process can be expressed simply in terms of the birth and death rates. By (13)

$$-q_{xx} = q_x = q_{x,x+1} + q_{x,x-1},$$

so that

(17) $$q_{xx} = -(\lambda_x + \mu_x) \qquad \text{and} \qquad q_x = \lambda_x + \mu_x.$$

Thus x is an absorbing state if and only if $\lambda_x = \mu_x = 0$. If x is a non-absorbing state, then by (12)

$$(18) \qquad Q_{xy} = \begin{cases} \dfrac{\mu_x}{\lambda_x + \mu_x}, & y = x - 1, \\ \dfrac{\lambda_x}{\lambda_x + \mu_x}, & y = x + 1, \\ 0, & \text{elsewhere.} \end{cases}$$

A birth and death process is called a *pure birth process* if $\mu_x = 0$, $x \in \mathcal{S}$, and a *pure death process* if $\lambda_x = 0$, $x \in \mathcal{S}$. A pure birth process can move only to the right, and a pure death process can move only to the left.

Given nonnegative numbers λ_x, $x \in \mathcal{S}$, and μ_x, $x \in \mathcal{S}$, it is natural to ask whether there is a birth and death process corresponding to these parameters. Of course, $\mu_0 = 0$ is a necessary requirement, as is $\lambda_d = 0$ if $\mathcal{S}$ is finite. The only additional problem is that explosions must be ruled out if $\mathcal{S}$ is infinite. It is not difficult to derive a necessary and sufficient condition for the process to be non-explosive. A simple sufficient condition for the process to be non-explosive is that for some positive numbers A and B

$$\lambda_x \leq A + Bx, \qquad x \geq 0.$$

This condition holds in all the examples we will consider.

In finding the birth and death rates of specific processes, we will use some standard properties of independent exponentially distributed random variables. Let $\xi_1, \ldots, \xi_n$ be independent random variables having exponential distributions with respective parameters $\alpha_1, \ldots, \alpha_n$. Then $\min(\xi_1, \ldots, \xi_n)$ has an exponential distribution with parameter $\alpha_1 + \cdots + \alpha_n$ and

$$(19) \qquad P(\xi_k = \min(\xi_1, \ldots, \xi_n)) = \frac{\alpha_k}{\alpha_1 + \cdots + \alpha_n}, \qquad k = 1, \ldots, n.$$

Moreover, with probability one, the random variables $\xi_1, \ldots, \xi_n$ take on n distinct values.

To verify these results we observe first that

$$\begin{aligned} P(\min(\xi_1, \ldots, \xi_n) > t) &= P(\xi_1 > t, \ldots, \xi_n > t) \\ &= P(\xi_1 > t) \cdots P(\xi_n > t) \\ &= e^{-\alpha_1 t} \cdots e^{-\alpha_n t} \\ &= e^{-(\alpha_1 + \cdots + \alpha_n)t}, \end{aligned}$$

and hence that $\min(\xi_1, \ldots, \xi_n)$ has the indicated exponential distribution.

3.2. Birth and death processes

Set
$$\eta_k = \min(\xi_j : j \neq k).$$

Then η_k has an exponential distribution with parameter
$$\beta_k = \sum_{j \neq k} \alpha_j,$$

and ξ_k and η_k are independent. Thus
$$\begin{aligned}
P(\xi_k = \min(\xi_1, \ldots, \xi_n)) &= P(\xi_k \leq \eta_k) \\
&= \int_0^\infty \left(\int_x^\infty \alpha_k e^{-\alpha_k x} \beta_k e^{-\beta_k y} \, dy \right) dx \\
&= \int_0^\infty \alpha_k e^{-\alpha_k x} e^{-\beta_k x} \, dx \\
&= \frac{\alpha_k}{\alpha_k + \beta_k} = \frac{\alpha_k}{\alpha_1 + \cdots + \alpha_n}.
\end{aligned}$$

In order to show that the random variables $\xi_1, \ldots, \xi_n$ take on n distinct values with probability one, it is enough to show that $P(\xi_i \neq \xi_j) = 1$ for $i \neq j$. But since ξ_i and ξ_j have a joint density f, it follows that
$$P(\xi_i = \xi_j) = \iint_{\{(x,y):\, x=y\}} f(x, y) \, dx \, dy = 0,$$

as desired.

Example 1. Branching process. Consider a collection of particles which act independently in giving rise to succeeding generations of particles. Suppose that each particle, from the time it appears, waits a random length of time having an exponential distribution with parameter q and then splits into two identical particles with probability p and disappears with probability $1 - p$. Let $X(t)$, $0 \leq t < \infty$, denote the number of particles present at time t. This branching process is a birth and death process. Find the birth and death rates.

Consider a branching process starting out with x particles. Let $\xi_1, \ldots, \xi_x$ be the times until these particles split apart or disappear. Then $\xi_1, \ldots, \xi_x$ each has an exponential distribution with parameter q, and hence $\tau_1 = \min(\xi_1, \ldots, \xi_x)$ has an exponential distribution with parameter $q_x = xq$. Whichever particle acts first has probability p of splitting into two particles and probability $1 - p$ of disappearing. Thus for $x \geq 1$
$$Q_{x,x+1} = p \quad \text{and} \quad Q_{x,x-1} = 1 - p.$$

State 0 is an absorbing state. Since $\lambda_x = q_x Q_{x,x+1}$ and $\mu_x = q_x Q_{x,x-1}$, we conclude that

$$\lambda_x = xqp \quad \text{and} \quad \mu_x = xq(1-p), \quad x \geq 0.$$

In the preceding example we did not actually prove that the process is a birth and death process, i.e., that it "starts from scratch" after making a jump. This intuitively reasonable property basically depends on the fact that an exponentially distributed random variable ξ satisfies the formula

$$P(\xi > t + s \mid \xi > s) = P(\xi > t), \quad s, t \geq 0,$$

but a rigorous proof is complicated.

By (17) and the definition of λ_x and μ_x, the backward and forward equations for a birth and death process can be written respectively as

(20) $\quad P'_{xy}(t) = \mu_x P_{x-1,y}(t) - (\lambda_x + \mu_x) P_{xy}(t) + \lambda_x P_{x+1,y}(t), \quad t \geq 0,$

and

(21) $\quad P'_{xy}(t) = \lambda_{y-1} P_{x,y-1}(t) - (\lambda_y + \mu_y) P_{xy}(t) + \mu_{y+1} P_{x,y+1}(t),$
$$t \geq 0.$$

In (21) we set $\lambda_{-1} = 0$, and if $\mathcal{S} = \{0, \ldots, d\}$ for $d < \infty$, we set $\mu_{d+1} = 0$.

We will solve the backward and forward equations for a birth and death process in some special cases. To do so we will use the result that if

(22) $\quad f'(t) = -\alpha f(t) + g(t), \quad t \geq 0,$

then

(23) $\quad f(t) = f(0)e^{-\alpha t} + \int_0^t e^{-\alpha(t-s)} g(s) \, ds, \quad t \geq 0.$

The proof of this standard result is very easy. We multiply (22) through by $e^{\alpha t}$ and rewrite the resulting equation as

$$\frac{d}{dt}(e^{\alpha t} f(t)) = e^{\alpha t} g(t).$$

Integrating from 0 to t we find that

$$e^{\alpha t} f(t) - f(0) = \int_0^t e^{\alpha s} g(s) \, ds,$$

and hence that (23) holds.

3.2.1. Two-state birth and death process.

Consider a birth and death process having state space $\mathcal{S} = \{0, 1\}$, and suppose that 0 and 1 are both non-absorbing states. Since $\mu_0 = \lambda_1 = 0$, the process is

3.2. Birth and death processes

determined by the parameters λ_0 and μ_1. For simplicity in notation we set $\lambda = \lambda_0$ and $\mu = \mu_1$. We can interpret such a process by thinking of state 1 as the system (e.g., telephone or machine) operating and state 0 as the system being idle. We suppose that starting from an idle state the system remains idle for a random length of time which is exponentially distributed with parameter λ, and that starting in an operating state the system continues operating for a random length of time which is exponentially distributed with parameter μ.

We will find the transition function of the process by solving the backward equation. It is left as an exercise for the reader to obtain the same results by solving the forward equation.

Setting $y = 0$ in (20), we see that

$$P'_{00}(t) = -\lambda P_{00}(t) + \lambda P_{10}(t), \qquad t \geq 0, \tag{24}$$

and

$$P'_{10}(t) = \mu P_{00}(t) - \mu P_{10}(t), \qquad t \geq 0. \tag{25}$$

Subtracting the second equation from the first,

$$\frac{d}{dt}(P_{00}(t) - P_{10}(t)) = -(\lambda + \mu)(P_{00}(t) - P_{10}(t)).$$

Applying (23),

$$P_{00}(t) - P_{10}(t) = (P_{00}(0) - P_{10}(0))e^{-(\lambda+\mu)t} \tag{26}$$
$$= e^{-(\lambda+\mu)t}.$$

Here we have used the formulas $P_{00}(0) = 1$ and $P_{10}(0) = 0$. It now follows from (24) that

$$P'_{00}(t) = -\lambda(P_{00}(t) - P_{10}(t))$$
$$= -\lambda e^{-(\lambda+\mu)t}.$$

Thus

$$P_{00}(t) = P_{00}(0) + \int_0^t P'_{00}(s)\, ds$$
$$= 1 - \int_0^t \lambda e^{-(\lambda+\mu)s}\, ds$$
$$= 1 - \frac{\lambda}{\lambda + \mu}(1 - e^{-(\lambda+\mu)t}),$$

or equivalently,

$$P_{00}(t) = \frac{\mu}{\lambda + \mu} + \frac{\lambda}{\lambda + \mu} e^{-(\lambda+\mu)t}, \qquad t \geq 0. \tag{27}$$

Now, by (26), $P_{10}(t) = P_{00}(t) - e^{-(\lambda+\mu)t}$, and therefore

(28) $$P_{10}(t) = \frac{\mu}{\lambda+\mu} - \frac{\mu}{\lambda+\mu} e^{-(\lambda+\mu)t}, \qquad t \geq 0.$$

By setting $y = 1$ in the backward equation, or by subtracting $P_{00}(t)$ and $P_{10}(t)$ from one, we conclude that

(29) $$P_{01}(t) = \frac{\lambda}{\lambda+\mu} - \frac{\lambda}{\lambda+\mu} e^{-(\lambda+\mu)t}, \qquad t \geq 0,$$

and

(30) $$P_{11}(t) = \frac{\lambda}{\lambda+\mu} + \frac{\mu}{\lambda+\mu} e^{-(\lambda+\mu)t}, \qquad t \geq 0.$$

From (27)–(30) we see that

(31) $$\lim_{t \to +\infty} P_{xy}(t) = \pi(y),$$

where

(32) $$\pi(0) = \frac{\mu}{\lambda+\mu} \quad \text{and} \quad \pi(1) = \frac{\lambda}{\lambda+\mu}.$$

If π_0 is the initial distribution of the process, then by (27) and (28)

$$P(X(t) = 0) = \pi_0(0)P_{00}(t) + (1 - \pi_0(0))P_{10}(t)$$
$$= \frac{\mu}{\lambda+\mu} + \left(\pi_0(0) - \frac{\mu}{\lambda+\mu}\right) e^{-(\lambda+\mu)t}, \qquad t \geq 0.$$

Similarly,

$$P(X(t) = 1) = \frac{\lambda}{\lambda+\mu} + \left(\pi_0(1) - \frac{\lambda}{\lambda+\mu}\right) e^{-(\lambda+\mu)t}, \qquad t \geq 0.$$

Thus $P(X(t) = 0)$ and $P(X(t) = 1)$ are independent of t if and only if π_0 is the distribution π given by (32).

3.2.2. Poisson process. Consider a pure birth process $X(t)$, $0 \leq t < \infty$, on the nonnegative integers such that

$$\lambda_x = \lambda > 0, \qquad x \geq 0.$$

Since a pure birth process can move only to the right,

(33) $$P_{xy}(t) = 0, \qquad y < x \text{ and } t \geq 0.$$

Also $P_{xx}(t) = P_x(\tau_1 > t)$ and hence

(34) $$P_{xx}(t) = e^{-\lambda t}, \qquad t \geq 0.$$

3.2. Birth and death processes

The forward equation for $y \neq 0$ is

$$P'_{xy}(t) = \lambda P_{x,y-1}(t) - \lambda P_{xy}(t), \qquad t \geq 0.$$

From (23) we see that

$$P_{xy}(t) = e^{-\lambda t} P_{xy}(0) + \lambda \int_0^t e^{-\lambda(t-s)} P_{x,y-1}(s)\, ds, \qquad t \geq 0.$$

Since $P_{xy}(0) = \delta_{xy}$, we conclude that for $y > x$

(35) $$P_{xy}(t) = \lambda \int_0^t e^{-\lambda(t-s)} P_{x,y-1}(s)\, ds, \qquad t \geq 0.$$

It follows from (34) and (35) that

$$P_{x,x+1}(t) = \lambda \int_0^t e^{-\lambda(t-s)} e^{-\lambda s}\, ds = \lambda e^{-\lambda t} \int_0^t ds = \lambda t e^{-\lambda t}$$

and hence by using (35) once more that

$$P_{x,x+2}(t) = \lambda \int_0^t e^{-\lambda(t-s)} \lambda s e^{-\lambda s}\, ds = \lambda^2 e^{-\lambda t} \int_0^t s\, ds = \frac{(\lambda t)^2}{2} e^{-\lambda t}.$$

By induction

(36) $$P_{xy}(t) = \frac{(\lambda t)^{y-x} e^{-\lambda t}}{(y-x)!}, \qquad 0 \leq x \leq y \text{ and } t \geq 0.$$

Formulas (33) and (36) imply that

(37) $$P_{xy}(t) = P_{0,y-x}(t), \qquad t \geq 0,$$

and that if $X(0) = x$, then $X(t) - x$ has a Poisson distribution with parameter λt.

In general, for $0 \leq s \leq t$, $X(t) - X(s)$ has a Poisson distribution with parameter $\lambda(t-s)$. For if $0 \leq s \leq t$ and y is a nonnegative integer, then

$$P(X(t) - X(s) = y) = \sum_x P(X(s) = x \text{ and } X(t) = x + y)$$

$$= \sum_x P(X(s) = x) P_{x,x+y}(t-s)$$

$$= \sum_x P(X(s) = x) P_{0y}(t-s)$$

$$= P_{0y}(t-s)$$

$$= \frac{(\lambda(t-s))^y e^{-\lambda(t-s)}}{y!}.$$

If $0 \leq t_1 \leq \cdots \leq t_n$, the random variables

$$X(t_2) - X(t_1), \ldots, X(t_n) - X(t_{n-1})$$

are independent. For we observe that if $z_1, \ldots, z_{n-1}$ are arbitrary integers, then by (6) and (37)

$$\begin{aligned}
P(X(t_2) - X(t_1) &= z_1, \ldots, X(t_n) - X(t_{n-1}) = z_{n-1}) \\
&= \sum_x P(X(t_1) = x) P_{0z_1}(t_2 - t_1) \cdots P_{0z_{n-1}}(t_n - t_{n-1}) \\
&= P_{0z_1}(t_2 - t_1) \cdots P_{0z_{n-1}}(t_n - t_{n-1}) \\
&= P(X(t_2) - X(t_1) = z_1) \cdots P(X(t_n) - X(t_{n-1}) = z_{n-1}).
\end{aligned}$$

By a *Poisson process with parameter* λ on $0 \leq t < \infty$, we mean a pure birth process $X(t)$, $0 \leq t < \infty$, having state space $\{0, 1, 2, \ldots\}$, constant birth rate $\lambda_x = \lambda > 0$, and initial value $X(0) = 0$. According to the above discussion the Poisson process satisfies the following three properties:

(i) $X(0) = 0$.
(ii) $X(t) - X(s)$ has a Poisson distribution with parameter $\lambda(t - s)$ for $0 \leq s \leq t$.
(iii) $X(t_2) - X(t_1), X(t_3) - X(t_2), \ldots, X(t_n) - X(t_{n-1})$ are independent for $0 \leq t_1 \leq t_2 \leq \cdots \leq t_n$.

The Poisson process can be used to model events occurring in time, such as calls coming into a telephone exchange, customers arriving at a queue, and radioactive disintegrations. Let $X(t)$, $0 \leq t < \infty$, denote the number of events occurring in the time interval $(0, t]$. For $0 \leq s \leq t$ the random variable $X(t) - X(s)$ denotes the number of events in the time interval $(s, t]$. If the waiting times between successive events are independent and exponentially distributed with common parameter λ, then $X(t)$, $0 \leq t < \infty$, is a Poisson process and properties (i)–(iii) hold. Property (ii) states that the number of events in any interval has a Poisson distribution. Property (iii) states that the numbers of events in disjoint time intervals are independent. Conversely, if $X(t)$, $0 \leq t < \infty$, satisfies properties (i)–(iii), then the waiting times between successive events are independent and exponentially distributed with common parameter λ, and hence $X(t)$ is a pure birth process with constant birth rate λ. This result was proved in Chapter 9 of Volume I, but will not be needed.

Since the Poisson process is a pure birth process starting in state 0, it follows that for $n \geq 1$ the time τ_n of the nth jump equals the time T_n when the process hits state n. When the Poisson process is used to model events occurring in time as described above, the common time $\tau_n = T_n$ is the time when the nth event occurs.

The Poisson process can be used to construct a variety of other processes.

3.2. Birth and death processes

Example 2. Branching process with immigration. Consider the branching process introduced in Example 1. Suppose that new particles immigrate into the system at random times that form a Poisson process with parameter λ and then give rise to succeeding generations as described in Example 1. Find the birth and death rates of this birth and death process.

Suppose there are initially x particles present. Let $\xi_1, \ldots, \xi_x$ be the times at which these particles split apart or disappear, and let η be the first time a new particle enters the system. We interpret the description of the system as implying that η is independent of $\xi_1, \ldots, \xi_x$. Then $\xi_1, \ldots, \xi_x$, η are independent exponentially distributed random variables having respective parameters $q, \ldots, q, \lambda$. Thus

$$\tau_1 = \min(\xi_1, \ldots, \xi_x, \eta)$$

is exponentially distributed with parameter $q_x = xq + \lambda$, and by (19)

$$P(\tau_1 = \eta) = \frac{\lambda}{xq + \lambda}.$$

The event $\{X(\tau_1) = x + 1\}$ occurs if either $\tau_1 = \eta$ or

$$\tau_1 = \min(\xi_1, \ldots, \xi_x)$$

and a particle splits into two new particles at time τ_1. Thus

$$Q_{x,x+1} = \frac{\lambda}{xq + \lambda} + \frac{xq}{xq + \lambda} p.$$

Also,

$$Q_{x,x-1} = \frac{xq}{xq + \lambda}(1 - p).$$

We conclude that

$$\lambda_x = q_x Q_{x,x+1} = xqp + \lambda$$

and

$$\mu_x = q_x Q_{x,x-1} = xq(1 - p).$$

It is also possible to construct a Poisson process with parameter λ on $-\infty < t < \infty$. We first construct two independent Poisson processes $X_1(t)$, $0 \le t < \infty$, and $X_2(t)$, $0 \le t < \infty$, both having parameter λ. We then define $X(t)$, $-\infty < t < \infty$, by

$$X(t) = \begin{cases} -X_1(-t), & t < 0, \\ X_2(t), & t \ge 0. \end{cases}$$

It is easy to show that the process $X(t)$, $-\infty < t < \infty$, so constructed, satisfies the following three properties:

(i) $X(0) = 0$.
(ii) $X(t) - X(s)$ has a Poisson distribution with parameter $\lambda(t - s)$ for $s \leq t$.
(iii) $X(t_2) - X(t_1), \ldots, X(t_n) - X(t_{n-1})$ are independent for $t_1 \leq t_2 \leq \cdots \leq t_n$.

3.2.3. Pure birth process. Consider a pure birth process $X(t)$, $0 \leq t < \infty$, on $\{0, 1, 2, \ldots\}$. The forward equation (21) reduces to

$$(38) \qquad P'_{xy}(t) = \lambda_{y-1} P_{x,y-1}(t) - \lambda_y P_{xy}(t), \qquad t \geq 0.$$

Since the process moves only to the right,

$$(39) \qquad P_{xy}(t) = 0, \qquad y < x \text{ and } t \geq 0.$$

It follows from (38) and (39) that

$$P'_{xx}(t) = -\lambda_x P_{xx}(t).$$

Since $P_{xx}(0) = 1$ and $P_{xy}(0) = 0$ for $y > x$, we conclude from (23) that

$$(40) \qquad P_{xx}(t) = e^{-\lambda_x t}, \qquad t \geq 0,$$

and

$$(41) \qquad P_{xy}(t) = \lambda_{y-1} \int_0^t e^{-\lambda_y(t-s)} P_{x,y-1}(s)\, ds, \qquad y > x \text{ and } t \geq 0.$$

We can use (40) and (41) to find $P_{xy}(t)$ recursively for $y > x$. In particular,

$$P_{x,x+1}(t) = \lambda_x \int_0^t e^{-\lambda_{x+1}(t-s)} e^{-\lambda_x s}\, ds,$$

and hence for $t \geq 0$

$$(42) \qquad P_{x,x+1}(t) = \begin{cases} \dfrac{\lambda_x}{\lambda_{x+1} - \lambda_x}(e^{-\lambda_x t} - e^{-\lambda_{x+1} t}), & \lambda_{x+1} \neq \lambda_x, \\ \lambda_x t e^{-\lambda_x t}, & \lambda_{x+1} = \lambda_x. \end{cases}$$

Example 3. *Linear birth process.* Consider a pure birth process on $\{0, 1, 2, \ldots\}$ having birth rates

$$\lambda_x = x\lambda, \qquad x \geq 0,$$

for some positive constant λ (the branching process with $p = 1$ is of this form). Find $P_{xy}(t)$.

As noted above, $P_{xy}(t) = 0$ for $y < x$ and

$$P_{xx}(t) = e^{-\lambda_x t} = e^{-x\lambda t}.$$

We see from (42) that

$$P_{x,x+1}(t) = xe^{-x\lambda t}(1 - e^{-\lambda t}).$$

To compute $P_{x,x+2}(t)$ we set $y = x + 2$ in (41) and obtain

$$P_{x,x+2}(t) = (x + 1)x\lambda \int_0^t e^{-(x+2)\lambda(t-s)}e^{-x\lambda s}(1 - e^{-\lambda s})\, ds$$

$$= (x + 1)x\lambda e^{-(x+2)\lambda t} \int_0^t e^{2\lambda s}(1 - e^{-\lambda s})\, ds$$

$$= (x + 1)x\lambda e^{-(x+2)\lambda t} \int_0^t e^{\lambda s}(e^{\lambda s} - 1)\, ds$$

$$= (x + 1)x\lambda e^{-(x+2)\lambda t} \frac{(e^{\lambda t} - 1)^2}{2\lambda}$$

$$= \binom{x+1}{2} e^{-x\lambda t}(1 - e^{-\lambda t})^2.$$

It is left as an exercise for the reader to show by induction that

(43) $$P_{xy}(t) = \binom{y-1}{y-x} e^{-x\lambda t}(1 - e^{-\lambda t})^{y-x}, \qquad y \geq x \text{ and } t \geq 0.$$

3.2.4. Infinite server queue.

Suppose that customers arrive for service according to a Poisson process with parameter λ and that each customer starts being served immediately upon his arrival (i.e., that there are an infinite number of servers). Suppose that the service times are independent and exponentially distributed with parameter μ. Let $X(t)$, $0 \leq t < \infty$, denote the number of customers in the process of being served at time t. This birth and death process, called an *infinite server queue*, is a special case of the branching process with immigration corresponding to $q = \mu$ and $p = 0$. We conclude that $\lambda_x = \lambda$ and $\mu_x = x\mu$, $x \geq 0$. The transition function $P_{xy}(t)$ will now be obtained by a probabilistic argument.

Let $Y(t)$ denote the number of customers who arrive in the time interval $(0, t]$. An interesting and useful result about the Poisson process is that conditioned on $Y(t) = k$, the times when the first k customers arrive are distributed as k independent random variables each uniformly distributed on $(0, t]$. In order to see intuitively why this should be true, consider an arbitrary partition $0 = t_0 < t_1 < \cdots < t_m = t$ of $[0, t]$ and let X_i denote the number of customers arriving between time t_{i-1} and time t_i. Then $X_1, \ldots, X_m$ are independent random variables having Poisson distributions with respective parameters

$$\lambda(t_1 - t_0), \ldots, \lambda(t_m - t_{m-1}),$$

and $X_1 + \cdots + X_m = Y(t)$ has a Poisson distribution with parameter λt. Thus for $x_1, \ldots, x_m$ nonnegative integers adding up to k,

$$P(X_1 = x_1, \ldots, X_m = x_m \mid Y(t) = k)$$
$$= P(X_1 = x_1, \ldots, X_m = x_m \mid X_1 + \cdots + X_m = k)$$
$$= \frac{P(X_1 = x_1, \ldots, X_m = x_m, X_1 + \cdots + X_m = k)}{P(X_1 + \cdots + X_m = k)}$$
$$= \frac{P(X_1 = x_1, \ldots, X_m = x_m)}{P(X_1 + \cdots + X_m = k)}$$
$$= \frac{\prod_{i=1}^{m} \dfrac{[\lambda(t_i - t_{i-1})]^{x_i} e^{-\lambda(t_i - t_{i-1})}}{x_i!}}{\dfrac{(\lambda t)^k e^{-\lambda t}}{k!}}$$
$$= \frac{k!}{\prod_{i=1}^{m} x_i!} \prod_{i=1}^{m} \left(\frac{t_i - t_{i-1}}{t}\right)^{x_i}.$$

But these multinomial probabilities are just those that would be obtained by choosing the k arrival times independently and uniformly distributed over $(0, t]$.

If a customer arrives at time $s \in (0, t]$, the probability that he is still in the process of being served at time t is $e^{-\mu(t-s)}$. Thus if a customer arrives at a time chosen uniformly from $(0, t]$, the probability that he is still in the process of being served at time t is

$$p_t = \frac{1}{t} \int_0^t e^{-\mu(t-s)} \, ds = \frac{1 - e^{-\mu t}}{\mu t}.$$

Let $X_1(t)$ denote the number of customers arriving in $(0, t]$ that are still in the process of being served at time t. It follows from the results of the previous two paragraphs that the conditional distribution of $X_1(t)$ given that $Y(t) = k$ is a binomial distribution with parameters k and p_t, i.e., that

$$P(X_1(t) = n \mid Y(t) = k) = \binom{k}{n} p_t^n (1 - p_t)^{k-n}.$$

3.2. Birth and death processes

Since $Y(t)$ has a Poisson distribution with parameter λt, we conclude that

$$P(X_1(t) = n) = \sum_{k=n}^{\infty} P(Y(t) = k, X_1(t) = n)$$

$$= \sum_{k=n}^{\infty} P(Y(t) = k) P(X_1(t) = n \mid Y(t) = k)$$

$$= \sum_{k=n}^{\infty} \frac{(\lambda t)^k e^{-\lambda t}}{k!} \cdot \frac{k!}{n!(k-n)!} p_t^n (1-p_t)^{k-n}$$

$$= \frac{(\lambda t p_t)^n e^{-\lambda t}}{n!} \sum_{k=n}^{\infty} \frac{(\lambda t(1-p_t))^{k-n}}{(k-n)!}$$

$$= \frac{(\lambda t p_t)^n e^{-\lambda t}}{n!} \sum_{m=0}^{\infty} \frac{(\lambda t(1-p_t))^m}{m!}$$

$$= \frac{(\lambda t p_t)^n e^{-\lambda t}}{n!} e^{\lambda t(1-p_t)}$$

$$= \frac{(\lambda t p_t)^n e^{-\lambda t p_t}}{n!}.$$

Thus $X_1(t)$ has a Poisson distribution with parameter

$$\lambda t p_t = \frac{\lambda}{\mu}(1 - e^{-\mu t}).$$

Let x denote the number of customers present initially and let $X_2(t)$ denote the number of these customers still in the process of being served at time t. Then $X_2(t)$ is independent of $X_1(t)$ and has a binomial distribution with parameters x and $e^{-\mu t}$. Since $X(t) = X_1(t) + X_2(t)$, we conclude that

$$P_{xy}(t) = P_x(X(t) = y) = \sum_{k=0}^{\min(x,y)} P_x(X_2(t) = k) P(X_1(t) = y - k).$$

Therefore

(44) $$P_{xy}(t) = \sum_{k=0}^{\min(x,y)} \left[\binom{x}{k} e^{-k\mu t}(1 - e^{-\mu t})^{x-k} \right.$$

$$\left. \times \frac{\left(\frac{\lambda}{\mu}(1 - e^{-\mu t})\right)^{y-k}}{(y-k)!} \exp\left(-\frac{\lambda}{\mu}(1 - e^{-\mu t})\right) \right].$$

As $t \to \infty$, $e^{-\mu t} \to 0$, and hence the terms in the sum in (44) all approach 0 except the term corresponding to $k = 0$. Consequently

(45) $$\lim_{t \to \infty} P_{xy}(t) = \frac{(\lambda/\mu)^y e^{-\lambda/\mu}}{y!}.$$

3.3. Properties of a Markov pure jump process

In this section we will discuss the notions of recurrence, transience, irreducibility, stationary distributions, and positive recurrence of Markov pure jump processes. The results will be described briefly and without proofs, as they are very similar to those for the Markov chains discussed in Chapters 1 and 2. In Section 3.3.1 we apply these results to birth and death processes.

Let $X(t)$, $0 \leq t < \infty$, be a Markov pure jump process having state space $\mathscr{S}$. For $y \in \mathscr{S}$ and $X(0) \neq y$, the first visit to y takes place at time

$$T_y = \min\,(t \geq 0 : X(t) = y).$$

If $X(0) = y$, then $\min\,(t \geq 0 : X(t) = y) = 0$. A more useful random variable in this case is the time T_y of the first return to y after the process leaves y. Both cases are covered by setting

$$T_y = \min\,(t \geq \tau_1 : X(t) = y).$$

Here τ_1 is the time of the first jump. If $\tau_1 = \infty$ or $X(t) \neq y$ for all $t \geq \tau_1$, we set $T_y = \infty$.

If x is an absorbing state, set $\rho_{xy} = \delta_{xy}$; and if x is a non-absorbing state, set

$$\rho_{xy} = P_x(T_y < \infty).$$

A state $y \in \mathscr{S}$ is called *recurrent* if $\rho_{yy} = 1$ and *transient* if $\rho_{yy} < 1$. The process is said to be a *recurrent process* if all of its states are recurrent and a *transient process* if all of its states are transient. The process is called *irreducible* if $\rho_{xy} > 0$ for all choices of $x \in \mathscr{S}$ and $y \in \mathscr{S}$.

The function $P(x, y) = Q_{xy}$, $x \in \mathscr{S}$ and $y \in \mathscr{S}$, is the transition function of a Markov chain called the *embedded chain*. The quantities ρ_{xy} for this Markov chain are equal to the corresponding quantities for the Markov pure jump process. This relationship shows that results of Chapter 1 involving only the numbers ρ_{xy} are also valid in the present context. In particular, an irreducible process is either a recurrent process or a transient process. It is recurrent if and only if the embedded chain is recurrent.

If $\pi(x)$, $x \in \mathscr{S}$, are nonnegative numbers summing to one and if

(46) $$\sum_x \pi(x) P_{xy}(t) = \pi(y), \qquad y \in \mathscr{S} \text{ and } t \geq 0,$$

then π is called a *stationary distribution*. If $X(0)$ has a stationary distribution π for its initial distribution, then

$$P(X(t) = y) = \sum_x \pi(x) P_{xy}(t) = \pi(y),$$

so that $X(t)$ has distribution π for all $t \geq 0$.

3.3. Properties of a Markov pure jump process

If (46) holds and $\mathscr{S}$ is finite, we can differentiate this equation and obtain

(47) $$\sum_x \pi(x) P'_{xy}(t) = 0, \qquad y \in \mathscr{S} \text{ and } t \geq 0.$$

In particular, by setting $t = 0$ in (47), we conclude from (11) that

(48) $$\sum_x \pi(x) q_{xy} = 0, \qquad y \in \mathscr{S}.$$

It can be shown that (47) and (48) are valid even if $\mathscr{S}$ is an infinite set. Suppose, conversely, that (48) holds. If $\mathscr{S}$ is finite we conclude from the backward equation (14) that

$$\frac{d}{dt} \sum_x \pi(x) P_{xy}(t) = \sum_x \pi(x) P'_{xy}(t)$$
$$= \sum_x \pi(x) \left(\sum_z q_{xz} P_{zy}(t) \right)$$
$$= \sum_z \left(\sum_x \pi(x) q_{xz} \right) P_{zy}(t)$$
$$= 0.$$

Thus

$$\sum_x \pi(x) P_{xy}(t)$$

is a constant in t and the constant value is given by

$$\sum_x \pi(x) P_{xy}(0) = \sum_x \pi(x) \delta_{xy} = \pi(y).$$

Consequently (46) holds. This conclusion is also valid if $\mathscr{S}$ is infinite, but the proof is much more complicated. In summary, (46) holds if and only if (48) holds.

A non-absorbing recurrent state x is called *positive recurrent* or *null recurrent* according as the *mean return time* $m_x = E_x(T_x)$ is finite or infinite. An absorbing state is considered to be positive recurrent. The process is said to be a *positive recurrent process* if all its states are positive recurrent and a *null recurrent process* if all its states are null recurrent. An irreducible recurrent process must be either a null recurrent process or a positive recurrent process. It can be shown that a stationary distribution is concentrated on the positive recurrent states, and hence a process that is transient or null recurrent has no stationary distribution. An irreducible positive recurrent process has a unique stationary distribution π, which, unless $\mathscr{S}$ consists of a single necessarily absorbing state, is given by

(49) $$\pi(x) = \frac{1}{q_x m_x}, \qquad x \in \mathscr{S}.$$

Formula (49) is intuitively reasonable. For in a large time interval $[0, t]$, the process makes about t/m_x visits to x and the average time in x per visit is $1/q_x$. Thus the total time spent in state x during the time interval $[0, t]$ should be about $t/(q_x m_x)$ and the proportion of time spent in state x should be about $1/(q_x m_x)$. This argument can be made rigorous by using the strong law of large numbers as was done in Section 2.3.

Markov pure jump processes do not have any periodicities, and, in particular, for an irreducible positive recurrent process having stationary distribution π,

$$\lim_{t \to \infty} P_{xy}(t) = \pi(y), \qquad x, y \in \mathcal{S}. \tag{50}$$

If $X(0)$ has the initial distribution $\pi_0(x)$, $x \in \mathcal{S}$, then

$$P(X(t) = y) = \sum_x \pi_0(x) P_{xy}(t),$$

which, by (50) and the bounded convergence theorem, converges to

$$\sum_x \pi_0(x) \pi(y) = \pi(y)$$

as $t \to \infty$. In other words

$$\lim_{t \to \infty} P(X(t) = y) = \pi(y),$$

and hence the distribution of $X(t)$ converges to the stationary distribution π regardless of the initial distribution of the process.

3.3.1. Applications to birth and death processes.

Let $X(t)$, $0 \le t < \infty$, be an irreducible birth and death process on $\{0, 1, 2, \ldots\}$. The process is transient if and only if the embedded birth and death chain having transition function $P(x, y) = Q_{xy}$, $x \ge 0$ and $y \ge 0$, is transient. From (18) in this chapter and the results in Section 1.7, we conclude that the birth and death process is transient if and only if

$$\sum_{x=1}^{\infty} \frac{\mu_1 \cdots \mu_x}{\lambda_1 \cdots \lambda_x} < \infty. \tag{51}$$

Equation (48) for a stationary distribution π becomes

$$\pi(1)\mu_1 - \pi(0)\lambda_0 = 0,$$

(52)

$$\pi(y+1)\mu_{y+1} - \pi(y)\lambda_y = \pi(y)\mu_y - \pi(y-1)\lambda_{y-1}, \qquad y \ge 1.$$

It follows easily from (52) and induction that

$$\pi(y+1)\mu_{y+1} - \pi(y)\lambda_y = 0, \qquad y \ge 0,$$

and hence that

$$\pi(y+1) = \frac{\lambda_y}{\mu_{y+1}} \pi(y), \qquad y \geq 0.$$

Consequently,

(53) $$\pi(x) = \frac{\lambda_0 \cdots \lambda_{x-1}}{\mu_1 \cdots \mu_x} \pi(0), \qquad x \geq 1.$$

Set

(54) $$\pi_x = \begin{cases} 1, & x = 0, \\ \frac{\lambda_0 \cdots \lambda_{x-1}}{\mu_1 \cdots \mu_x}, & x \geq 1. \end{cases}$$

Then (53) can be written as

(55) $$\pi(x) = \pi_x \pi(0), \qquad x \geq 0.$$

Conversely, (52) follows from (54) and (55).

Suppose now that $\sum_x \pi_x < \infty$, i.e., that

(56) $$\sum_{x=1}^{\infty} \frac{\lambda_0 \cdots \lambda_{x-1}}{\mu_1 \cdots \mu_x} < \infty.$$

We conclude from (55) that the birth and death process has a unique stationary distribution π, given by

(57) $$\pi(x) = \frac{\pi_x}{\sum_{y=0}^{\infty} \pi_y}, \qquad x \geq 0.$$

If (56) fails to hold, the birth and death process has no stationary distribution.

In summary, an irreducible birth and death process on $\{0, 1, 2, \ldots\}$ is transient if and only if (51) holds, positive recurrent if and only if (56) holds, and null recurrent if and only if (51) and (56) each fail to hold, i.e., if and only if

(58) $$\sum_{x=1}^{\infty} \frac{\mu_1 \cdots \mu_x}{\lambda_1 \cdots \lambda_x} = \infty \quad \text{and} \quad \sum_{x=1}^{\infty} \frac{\lambda_0 \cdots \lambda_{x-1}}{\mu_1 \cdots \mu_x} = \infty.$$

An irreducible birth and death process having finite state space $\{0, 1, \ldots, d\}$ is necessarily positive recurrent. It has a unique stationary distribution given by

(59) $$\pi(x) = \frac{\pi_x}{\sum_{y=0}^{d} \pi_y}, \qquad 0 \leq x \leq d,$$

where π_x, $0 \leq x \leq d$, is given by (54).

Example 4. Show that the infinite server queue is positive recurrent and find its stationary distribution.

The infinite server queue has state space $\{0, 1, 2, \ldots\}$ and birth and death rates
$$\lambda_x = \lambda \quad \text{and} \quad \mu_x = x\mu, \quad x \geq 0.$$
This process is clearly irreducible. It follows from (54) that
$$\pi_x = \frac{\lambda^x}{x!\,\mu^x} = \frac{(\lambda/\mu)^x}{x!}, \quad x \geq 0.$$
Since
$$\sum_{x=0}^{\infty} \frac{(\lambda/\mu)^x}{x!} = e^{\lambda/\mu}$$
is finite, we conclude that the process is positive recurrent and has the unique stationary distribution π given by
$$(60) \qquad \pi(x) = \frac{(\lambda/\mu)^x}{x!} e^{-\lambda/\mu}, \quad x \geq 0,$$
which we note is a Poisson distribution with parameter λ/μ. We also note that (50) holds for this process, a direct consequence of (45) and (60).

Example 5. *N server queue.* Suppose customers arrive according to a Poisson process with parameter $\lambda > 0$. They are served by N servers, where N is a finite positive number. Suppose the service times are exponentially distributed with parameter μ and that whenever there are more than N customers waiting for service the excess customers form a queue and wait until their turn at one of the N servers. This process is a birth and death process on $\{0, 1, 2, \ldots\}$ with birth rates $\lambda_x = \lambda$, $x \geq 0$, and death rates
$$\mu_x = \begin{cases} x\mu, & 0 \leq x < N, \\ N\mu, & x \geq N. \end{cases}$$
Determine when this process is transient, null recurrent, and positive recurrent; and find the stationary distribution in the positive recurrent case.

Condition (51) for transience reduces to
$$\sum_{x=0}^{\infty} \left(\frac{N\mu}{\lambda}\right)^x < \infty.$$

Thus the N server queue is transient if and only if $N\mu < \lambda$. Condition (56) for positive recurrence reduces to

$$\sum_{x=0}^{\infty} \left(\frac{\lambda}{N\mu}\right)^x < \infty.$$

The N server queue is therefore positive recurrent if and only if $\lambda < N\mu$. Consequently the N server queue is null recurrent if and only if $\lambda = N\mu$. These results naturally are similar to those for the 1 server queue discussed in Chapters 1 and 2.

In the positive recurrent case,

$$\pi_x = \begin{cases} \dfrac{(\lambda/\mu)^x}{x!}, & 0 \leq x < N, \\ \dfrac{(\lambda/\mu)^x}{N! \, N^{x-N}}, & x \geq N. \end{cases}$$

Set

$$K = \sum_{x=0}^{\infty} \pi_x = \sum_{x=0}^{N-1} \frac{(\lambda/\mu)^x}{x!} + \frac{(\lambda/\mu)^N}{N!}\left(1 - \frac{\lambda}{N\mu}\right)^{-1}.$$

We conclude that if $\lambda < N\mu$, the stationary distribution is given by

$$\pi(x) = \begin{cases} \dfrac{1}{K}\dfrac{(\lambda/\mu)^x}{x!}, & 0 \leq x < N, \\ \dfrac{1}{K}\dfrac{(\lambda/\mu)^x}{N! \, N^{x-N}}, & x \geq N. \end{cases}$$

Exercises

1 Find the transition function of the two-state birth and death process by solving the forward equation.

2 Consider a birth and death process having three states 0, 1, and 2, and birth and death rates such that $\lambda_0 = \mu_2$. Use the forward equation to find $P_{0y}(t)$, $y = 0, 1, 2$.

Exercises 3–8 all refer to events occurring in time according to a Poisson process with parameter λ on $0 \leq t < \infty$. Here $X(t)$ denotes the number of events that occur in the time interval $(0, t]$.

3 Find the conditional probability that there are m events in the first s units of time, given that there are n events in the first t units of time, where $0 \leq m \leq n$ and $0 \leq s \leq t$.

4 Let T_m denote the time to the mth event. Find the distribution function of T_m. *Hint*: $\{T_m \leq t\} = \{X(t) \geq m\}$.

5 Find the density of the random variable T_m in Exercise 4. *Hint*: First consider some specific cases, say, $m = 1, 2, 3$.

6 Find $P(T_1 \leq s \mid X(t) = n)$ for $0 \leq s \leq t$ and n a positive integer.

7 Let T be a random variable that is independent of the times when events occur. Suppose that T has an exponential density with parameter v:

$$f_T(t) = \begin{cases} ve^{-vt}, & t > 0, \\ 0, & t \leq 0. \end{cases}$$

Find the distribution of $X(T)$, which is the number of events occurring by time T. *Hint:* Use the formulas

$$P(X(T) = n) = \int_0^\infty f_T(t) P(X(T) = n \mid T = t) \, dt$$

and

$$P(X(T) = n \mid T = t) = P(X(t) = n).$$

8 Solve the previous exercise if T has the gamma density with parameters α and v:

$$f_T(t) = \begin{cases} v^\alpha t^{\alpha-1} e^{-vt}/\Gamma(\alpha), & t > 0, \\ 0, & t \leq 0. \end{cases}$$

9 Verify Equation (43).

10 Consider a pure death process on $\{0, 1, 2, \ldots\}$.
  (a) Write the forward equation.
  (b) Find $P_{xx}(t)$.
  (c) Solve for $P_{xy}(t)$ in terms of $P_{x,y+1}(t)$.
  (d) Find $P_{x,x-1}(t)$.
  (e) Show that if $\mu_x = x\mu$, $x \geq 0$, for some constant μ, then

$$P_{xy}(t) = \binom{x}{y} (e^{-\mu t})^y (1 - e^{-\mu t})^{x-y}, \qquad 0 \leq y \leq x.$$

11 Let $X(t)$, $t \geq 0$, be the infinite server queue and suppose that initially there are x customers present. Compute the mean and variance of $X(t)$.

12 Consider a birth and death process $X(t)$, $t \geq 0$, such as the branching process, that has state space $\{0, 1, 2, \ldots\}$ and birth and death rates of the form

$$\lambda_x = x\lambda \qquad \text{and} \qquad \mu_x = x\mu, \qquad x \geq 0,$$

where λ and μ are nonnegative constants. Set

$$m_x(t) = E_x(X(t)) = \sum_{y=0}^\infty y P_{xy}(t).$$

  (a) Write the forward equation for the process.
  (b) Use the forward equation to show that $m_x'(t) = (\lambda - \mu) m_x(t)$.
  (c) Conclude that

$$m_x(t) = xe^{(\lambda-\mu)t}.$$

13 Let $X(t)$, $t > 0$, be as in Exercise 12. Set

$$s_x(t) = E_x(X^2(t)) = \sum_{y=0}^\infty y^2 P_{xy}(t).$$

(a) Use the forward equation to show that
$$s'_x(t) = 2(\lambda - \mu)s_x(t) + (\lambda + \mu)m_x(t).$$
(b) Find $s_x(t)$.
(c) Find Var $X(t)$ under the condition that $X(0) = x$.

14 Suppose d particles are distributed into two boxes. A particle in box 0 remains in that box for a random length of time that is exponentially distributed with parameter λ before going to box 1. A particle in box 1 remains there for an amount of time that is exponentially distributed with parameter μ before going to box 0. The particles act independently of each other. Let $X(t)$ denote the number of particles in box 1 at time $t \geq 0$. Then $X(t)$, $t \geq 0$, is a birth and death process on $\{0, \ldots, d\}$.
(a) Find the birth and death rates.
(b) Find $P_{xd}(t)$. Hint: Let $X_i(t)$, $i = 0$ or 1, denote the number of particles in box 1 at time $t \geq 0$ that started in box i at time 0, so that $X(t) = X_0(t) + X_1(t)$. If $X(0) = x$, then $X_0(t)$ and $X_1(t)$ are independent and binomially distributed with parameters defined in terms of x and the transition function of the two-state birth and death process.
(c) Find $E_x(X(t))$.

15 Consider the infinite server queue discussed in Section 3.2.4. Let $X_1(t)$ and $X_2(t)$ be as defined there. Suppose that the initial distribution π_0 is a Poisson distribution with parameter v.
(a) Use the formula
$$P(X_2(t) = k) = \sum_{x=k}^{\infty} \pi_0(x) P_x(X_2(t) = k)$$
to show that $X_2(t)$ has a Poisson distribution with parameter $ve^{-\mu t}$.
(b) Use the result of (a) to show that $X(t) = X_1(t) + X_2(t)$ has a Poisson distribution with parameter
$$\frac{\lambda}{\mu} + \left(v - \frac{\lambda}{\mu}\right) e^{-\mu t}.$$
(c) Conclude that $X(t)$ has the same distribution as $X(0)$ if and only if $v = \lambda/\mu$.

16 Consider a birth and death process on the nonnegative integers whose death rates are given by $\mu_x = x$, $x \geq 0$. Determine whether the process is transient, null recurrent, or positive recurrent if the birth rates are
(a) $\lambda_x = x + 1$, $x \geq 0$;
(b) $\lambda_x = x + 2$, $x \geq 0$.

17 Let $X(t)$, $t \geq 0$, be a birth and death process on the nonnegative integers such that $\lambda_x > 0$ and $\mu_x > 0$ for $x \geq 1$. Set $\gamma_0 = 1$ and

$$\gamma_x = \frac{\mu_1 \cdots \mu_x}{\lambda_1 \cdots \lambda_x}, \qquad x \geq 1.$$

(a) Show that if $\sum_{y=0}^{\infty} \gamma_y = \infty$, then $\rho_{x0} = 1$, $x \geq 1$.
(b) Show that if $\sum_{y=0}^{\infty} \gamma_y < \infty$, then

$$\rho_{x0} = \frac{\sum_{y=x}^{\infty} \gamma_y}{\sum_{y=0}^{\infty} \gamma_y}, \qquad x \geq 1.$$

Hint: Use Exercise 26 of Chapter 1.

18 Let $X(t)$, $t \geq 0$, be a single server queue ($N = 1$ in Example 5).
(a) Show that if $\mu \geq \lambda > 0$, then $\rho_{x0} = 1$, $x \geq 1$.
(b) Show that if $\mu < \lambda$, then

$$\rho_{x0} = (\mu/\lambda)^x, \qquad x \geq 1.$$

19 Consider the branching process introduced in Example 1. Use Exercise 17 to show that if $p \leq \frac{1}{2}$, then $\rho_{x0} = 1$ for all x and that if $p > \frac{1}{2}$, then

$$\rho_{x0} = \left(\frac{1-p}{p}\right)^x, \qquad x \geq 1.$$

20 Find the stationary distribution for the process in Exercise 14.

21 Suppose d machines are subject to failures and repairs. The failure times are exponentially distributed with parameter μ, and the repair times are exponentially distributed with parameter λ. Let $X(t)$ denote the number of machines that are in satisfactory order at time t. If there is only one repairman, then under appropriate reasonable assumptions, $X(t)$, $t \geq 0$, is a birth and death process on $\{0, 1, \ldots, d\}$ with birth rates $\lambda_x = \lambda$, $0 \leq x < d$, and death rates $\mu_x = x\mu$, $0 \leq x \leq d$. Find the stationary distribution for this process.

22 Consider a positive recurrent irreducible birth and death process on $\mathcal{S} = \{0, 1, 2, \ldots\}$, and let $X(0)$ have the stationary distribution π for its initial distribution. Then $X(t)$ has distribution π for all $t \geq 0$. The quantities

$$E\lambda_{X(t)} = \sum_{x=0}^{\infty} \lambda_x \pi(x) \quad \text{and} \quad E\mu_{X(t)} = \sum_{x=0}^{\infty} \mu_x \pi(x)$$

can be interpreted, respectively, as the average birth rate and the average death rate of the process.
(a) Show that the average birth rate equals the average death rate.
(b) What does (a) imply about a positive recurrent N server queue?

4

Second Order Processes

A *stochastic process* can be defined quite generally as any collection of random variables $X(t)$, $t \in T$, defined on a common probability space, where T is a subset of $(-\infty, \infty)$ and is usually thought of as the time parameter set. The process is called a *continuous parameter process* if T is an interval having positive length and a *discrete parameter process* if T is a subset of the integers. If $T = \{0, 1, 2, \ldots\}$ it is usual to denote the process by X_n, $n \geq 0$. The Markov chains discussed in Chapters 1 and 2 are discrete parameter processes, while the pure jump processes discussed in Chapter 3 are continuous parameter processes.

A stochastic process $X(t)$, $t \in T$, is called a *second order process* if $EX^2(t) < \infty$ for each $t \in T$. Second order processes and random variables defined in terms of them by various "linear" operations including integration and differentiation are the subjects of this and the next two chapters. We will obtain formulas for the means, variances, and covariances of such random variables.

We will consider continuous parameter processes almost exclusively in these three chapters. Since no new techniques are needed for handling the analogous results for discrete parameter processes, little would be gained by treating such processes in detail.

4.1. Mean and covariance functions

Let $X(t)$, $t \in T$, be a second order process. The *mean function* $\mu_x(t)$, $t \in T$, of the process is defined by

$$\mu_x(t) = EX(t).$$

The *covariance function* $r_x(s, t)$, $s \in T$ and $t \in T$, is defined by

$$r_x(s, t) = \text{cov}(X(s), X(t)) = EX(s)X(t) - EX(s)EX(t).$$

This function is also called the auto-covariance function to distinguish it from the cross-covariance function which will be defined later. Since

Var $X(t) = \operatorname{cov}(X(t), X(t))$, the variance of $X(t)$ can be expressed in terms of the covariance function as

(1) $$\operatorname{Var} X(t) = r_X(t, t), \qquad t \in T.$$

By a *finite linear combination* of the random variables $X(t)$, $t \in T$, we mean a random variable of the form

$$\sum_{j=1}^{n} b_j X(t_j),$$

where n is a positive integer, $t_1, \ldots, t_n$ are points in T, and $b_1, \ldots, b_n$ are real constants. The covariance between two such finite linear combinations is given by

$$\operatorname{cov}\left(\sum_{i=1}^{m} a_i X(s_i), \sum_{j=1}^{n} b_j X(t_j)\right) = \sum_{i=1}^{m} \sum_{j=1}^{n} a_i b_j \operatorname{cov}(X(s_i), X(t_j))$$
$$= \sum_{i=1}^{m} \sum_{j=1}^{n} a_i b_j r_X(s_i, t_j).$$

In particular,

(2) $$\operatorname{Var}\left(\sum_{j=1}^{n} b_j X(t_j)\right) = \sum_{i=1}^{n} \sum_{j=1}^{n} b_i b_j r_X(t_i, t_j).$$

It follows immediately from the definition of the covariance function that it is *symmetric* in s and t, i.e., that

(3) $$r_X(s, t) = r_X(t, s), \qquad s, t \in T.$$

It is also *nonnegative definite*. That is, if n is a positive integer, $t_1, \ldots, t_n$ are in T, and $b_1, \ldots, b_n$ are real numbers, then

$$\sum_{i=1}^{n} \sum_{j=1}^{n} b_i b_j r_X(t_i, t_j) \geq 0.$$

This is an immediate consequence of (2).

We say that $X(t)$, $-\infty < t < \infty$, is a *second order stationary process* if for every number τ the second order process $Y(t)$, $-\infty < t < \infty$, defined by

$$Y(t) = X(t + \tau), \qquad -\infty < t < \infty,$$

has the same mean and covariance functions as the $X(t)$ process. It is left as an exercise for the reader to show that this is the case if and only if $\mu_X(t)$ is independent of t and $r_X(s, t)$ depends only on the difference between s and t.

Let $X(t)$, $-\infty < t < \infty$, be a second order stationary process. Then

$$\mu_X(t) = \mu_X, \qquad -\infty < t < \infty,$$

4.1. Mean and covariance functions

where μ_X denotes the common mean of the random variables $X(t)$, $-\infty < t < \infty$. Since $r_X(s, t)$ depends only on the difference between s and t,

(4) $\qquad r_X(s, t) = r_X(0, t - s), \qquad -\infty < s, t < \infty.$

The function $r_X(t)$, $-\infty < t < \infty$, defined by

(5) $\qquad r_X(t) = r_X(0, t), \qquad -\infty < t < \infty,$

is also called the *covariance function* (or auto-covariance function) of the process. We see from (4) and (5) that

$$r_X(s, t) = r_X(t - s), \qquad -\infty < s, t < \infty.$$

It follows from (3) that $r_X(t)$ is symmetric about the origin, i.e., that

$$r_X(-t) = r_X(t), \qquad -\infty < t < \infty.$$

The random variables $X(t)$, $-\infty < t < \infty$, have a common variance given by

$$\operatorname{Var} X(t) = r_X(0), \qquad -\infty < t < \infty.$$

Recall Schwarz's inequality, which asserts that if X and Y are random variables having finite second moment, then $(EXY)^2 \leq EX^2 EY^2$. Applying Schwarz's inequality to the random variables $X - EX$ and $Y - EY$, we see that $(\operatorname{cov}(X, Y))^2 \leq \operatorname{Var} X \operatorname{Var} Y$.

It follows from this last inequality that

$$|\operatorname{cov}(X(0), X(t))| \leq \sqrt{\operatorname{Var} X(0) \operatorname{Var} X(t)},$$

and hence that

$$|r_X(t)| \leq r_X(0), \qquad -\infty < t < \infty.$$

If $r_X(0) > 0$, the correlation between $X(s)$ and $X(s + t)$ is given independently of s by

$$\frac{\operatorname{cov}(X(s), X(s + t))}{\sqrt{\operatorname{Var} X(s)} \sqrt{\operatorname{Var} X(t)}} = \frac{r_X(t)}{r_X(0)}, \qquad -\infty < s, t < \infty.$$

Example 1. Let Z_1 and Z_2 be independent normally distributed random variables each having mean 0 and variance σ^2. Let λ be a real constant and set $X(t) = Z_1 \cos \lambda t + Z_2 \sin \lambda t$, $-\infty < t < \infty$. Find the mean and covariance functions of $X(t)$, $-\infty < t < \infty$, and show that it is a second order stationary process.

We observe first that

$$\mu_X(t) = EZ_1 \cos \lambda t + EZ_2 \sin \lambda t = 0, \qquad -\infty < t < \infty.$$

Next,

$$\begin{aligned}
r_X(s, t) &= \text{cov}(X(s), X(t)) \\
&= EX(s)X(t) - EX(s)EX(t) \\
&= EX(s)X(t) \\
&= E(Z_1 \cos \lambda s + Z_2 \sin \lambda s)(Z_1 \cos \lambda t + Z_2 \sin \lambda t) \\
&= EZ_1^2 \cos \lambda s \cos \lambda t + EZ_2^2 \sin \lambda s \sin \lambda t \\
&= \sigma^2 (\cos \lambda s \cos \lambda t + \sin \lambda s \sin \lambda t) \\
&= \sigma^2 \cos \lambda (t - s).
\end{aligned}$$

This shows that $X(t)$, $-\infty < t < \infty$, is a second order stationary process having mean zero and covariance function

$$r_X(t) = \sigma^2 \cos \lambda t, \quad -\infty < t < \infty.$$

Example 2. Consider a two-state birth and death process as discussed in Section 3.2.1. It follows from that discussion that the transition probabilities of the process are given by

(6)
$$P_{00}(t) = 1 - P_{01}(t) = \frac{\mu}{\lambda + \mu} + \frac{\lambda}{\lambda + \mu} e^{-(\lambda + \mu)t}, \quad t \geq 0,$$

$$P_{11}(t) = 1 - P_{10}(t) = \frac{\lambda}{\lambda + \mu} + \frac{\mu}{\lambda + \mu} e^{-(\lambda + \mu)t}, \quad t \geq 0,$$

where λ and μ are positive constants. The process has the stationary distribution defined by

(7) $$\pi(0) = \frac{\mu}{\lambda + \mu} \quad \text{and} \quad \pi(1) = \frac{\lambda}{\lambda + \mu}.$$

In Chapter 3 we discussed birth and death processes defined on $0 \leq t < \infty$. Actually in the positive recurrent case it is possible to construct a corresponding process on $-\infty < t < \infty$ having the stationary distribution determined by (7). This process will be such that

(8) $$P(X(t) = 0) = \frac{\mu}{\lambda + \mu} \quad \text{and} \quad P(X(t) = 1) = \frac{\lambda}{\lambda + \mu},$$

$$-\infty < t < \infty,$$

and such that the Markov property

(9) $$P(X(t) = y \mid X(s) = x) = P_{xy}(t - s), \quad -\infty < s \leq t < \infty,$$

4.1. Mean and covariance functions

holds, where $P_{xy}(t)$, $t \geq 0$, is given by (6). We will show that such a process is a second order stationary process and find its mean and covariance functions.

The mean function is given by

$$\mu_X(t) = EX(t)$$
$$= 0 \cdot P(X(t) = 0) + 1 \cdot P(X(t) = 1) = \frac{\lambda}{\lambda + \mu}.$$

Let $-\infty < s \leq t < \infty$. Then

$$EX(s)X(t) = P(X(s) = 1 \text{ and } X(t) = 1)$$
$$= P(X(s) = 1)P(X(t) = 1 \mid X(s) = 1)$$
$$= P(X(s) = 1)P_{11}(t - s)$$
$$= \frac{\lambda}{\lambda + \mu}\left(\frac{\lambda}{\lambda + \mu} + \frac{\mu}{\lambda + \mu} e^{-(\lambda+\mu)(t-s)}\right)$$
$$= \left(\frac{\lambda}{\lambda + \mu}\right)^2 + \frac{\lambda\mu}{(\lambda + \mu)^2} e^{-(\lambda+\mu)(t-s)}.$$

It follows that

$$r_X(s, t) = \frac{\lambda\mu}{(\lambda + \mu)^2} e^{-(\lambda+\mu)(t-s)}, \quad -\infty < s \leq t < \infty.$$

By symmetry we see that

$$r_X(s, t) = \frac{\lambda\mu}{(\lambda + \mu)^2} e^{-(\lambda+\mu)|t-s|}, \quad -\infty < s, t < \infty.$$

Thus $X(t)$, $-\infty < t < \infty$, is a second order stationary process having mean $\lambda/(\lambda + \mu)$ and covariance function

$$r_X(t) = \frac{\lambda\mu}{(\lambda + \mu)^2} e^{-(\lambda+\mu)|t|}, \quad -\infty < t < \infty.$$

Other interesting examples of second order processes can be obtained from Poisson processes.

Example 3. Consider a Poisson process $X(t)$, $-\infty < t < \infty$, with parameter λ (see Section 3.2.2). This process satisfies the following properties:
 (i) $X(0) = 0$.
 (ii) $X(t) - X(s)$ has a Poisson distribution with mean $\lambda(t - s)$ for $s \leq t$.
 (iii) $X(t_2) - X(t_1), X(t_3) - X(t_2), \ldots, X(t_n) - X(t_{n-1})$ are independent for $t_1 \leq t_2 \leq \cdots \leq t_n$.

We will now find the mean and covariance function of a process $X(t)$, $-\infty < t < \infty$, satisfying (i)–(iii). It follows from properties (i) and (ii) that $X(t)$ has a Poisson distribution with mean λt for $t \geq 0$ and $-X(t)$ has a Poisson distribution with mean $\lambda(-t)$ for $t < 0$. Thus

$$\mu_X(t) = \lambda t, \qquad -\infty < t < \infty.$$

Since the variance of a Poisson distribution equals its mean, we see that $X(t)$ has finite second moment and that $\operatorname{Var} X(t) = \lambda|t|$. Let $0 \leq s \leq t$. Then

$$\operatorname{cov}(X(s), X(s)) = \operatorname{Var} X(s) = \lambda s.$$

It follows from properties (i) and (iii) that $X(s)$ and $X(t) - X(s)$ are independent, and hence

$$\operatorname{cov}(X(s), X(t) - X(s)) = 0.$$

Thus

$$\operatorname{cov}(X(s), X(t)) = \operatorname{cov}(X(s), X(s) + X(t) - X(s))$$
$$= \operatorname{cov}(X(s), X(s)) + \operatorname{cov}(X(s), X(t) - X(s))$$
$$= \lambda s.$$

If $s < 0$ and $t > 0$, then by properties (i) and (iii) the random variables $X(s)$ and $X(t)$ are independent, and hence

$$\operatorname{cov}(X(s), X(t)) = 0.$$

The other cases can be handled similarly. We find in general that

$$(10) \qquad r_X(s, t) = \begin{cases} \lambda \min(|s|, |t|), & st \geq 0, \\ 0, & st < 0. \end{cases}$$

The process from Example 3 is not a second order stationary process. In the next example we will consider a closely related process which is a second order stationary process.

Example 4. Let $X(t)$, $-\infty < t < \infty$, be a Poisson process with parameter λ. Set

$$Y(t) = X(t + 1) - X(t), \qquad -\infty < t < \infty.$$

Find the mean and covariance function of the $Y(t)$ process, and show that it is a second order stationary process.

Since $EX(t) = \lambda t$, it follows that

$$EY(t) = E(X(t + 1) - X(t))$$
$$= \lambda(t + 1) - \lambda t = \lambda,$$

so the random variables $Y(t)$ have common mean λ. To compute the covariance function of the $Y(t)$ process, we observe that if $|t - s| \geq 1$, then the random variables $X(s + 1) - X(s)$ and $X(t + 1) - X(t)$ are independent by property (iii). Consequently,

$$r_Y(s, t) = 0 \quad \text{for} \quad |t - s| \geq 1.$$

Suppose $s \leq t < s + 1$. Then

$$\begin{aligned}
\text{cov}\,(Y(s), Y(t)) &= \text{cov}\,(X(s + 1) - X(s), X(t + 1) - X(t)) \\
&= \text{cov}\,(X(t) - X(s) + X(s + 1) - X(t), X(s + 1) \\
&\quad - X(t) + X(t + 1) - X(s + 1)).
\end{aligned}$$

It follows from property (iii) and the assumptions on s and t that

$$\text{cov}\,(X(t) - X(s), X(s + 1) - X(t)) = 0,$$
$$\text{cov}\,(X(t) - X(s), X(t + 1) - X(s + 1)) = 0,$$

and

$$\text{cov}\,(X(s + 1) - X(t), X(t + 1) - X(s + 1)) = 0.$$

By property (ii)

$$\begin{aligned}
\text{cov}\,(X(s + 1) - X(t), X(s + 1) - X(t)) &= \text{Var}\,(X(s + 1) - X(t)) \\
&= \lambda(s + 1 - t).
\end{aligned}$$

Thus

$$\text{cov}\,(Y(s), Y(t)) = \lambda(s + 1 - t).$$

By using symmetry we find in general that

$$r_Y(s, t) = \begin{cases} \lambda(1 - |t - s|), & |t - s| < 1, \\ 0, & |t - s| \geq 1. \end{cases}$$

Thus $Y(t)$, $-\infty < t < \infty$, is a second order stationary process having mean λ and covariance function

$$r_Y(t) = \begin{cases} \lambda(1 - |t|), & |t| < 1, \\ 0, & |t| \geq 1. \end{cases}$$

In Figure 1 we have graphed the covariance function for three different second order stationary processes. These covariance functions are special cases of those found in Examples 1, 2, and 4 respectively. In each case $r_X(0) = 1$ and hence $r_X(t)$ is equal to the correlation between $X(0)$ and $X(t)$. In the top curve of Figure 1 we see that the correlation oscillates between -1 and 1. In the middle curve the correlation decreases exponentially fast as $|t| \to \infty$. In the bottom curve the correlation decreases linearly to zero as $|t|$ increases from 0 to 1 and remains zero for all larger values of $|t|$.

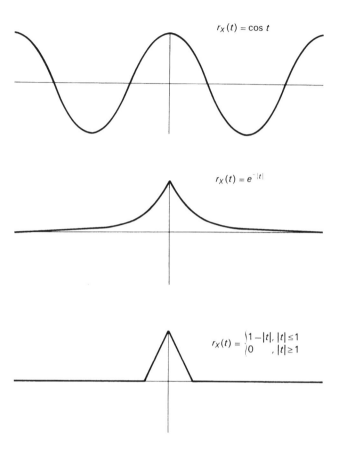

Figure 1

Consider two second order processes $X(t)$, $t \in T$, and $Y(t)$, $t \in T$. Their *cross-covariance function* is defined as

$$r_{XY}(s, t) = \text{cov}(X(s), Y(t)), \qquad s, t \in T.$$

Clearly

$$r_{XY}(s, t) = r_{YX}(t, s)$$

and

$$r_{XX}(s, t) = r_X(s, t).$$

The cross-covariance function can be used to find the covariance function of the sum of two processes. Indeed,

$$\begin{aligned} r_{X+Y}(s, t) &= \text{cov}(X(s) + Y(s), X(t) + Y(t)) \\ &= r_{XX}(s, t) + r_{XY}(s, t) + r_{YX}(s, t) + r_{YY}(s, t), \end{aligned}$$

which can be rewritten as

(11) $\qquad r_{X+Y}(s, t) = r_X(s, t) + r_{XY}(s, t) + r_{YX}(s, t) + r_Y(s, t).$

In the important case when the cross-covariance function vanishes, (11) reduces to

(12) $$r_{X+Y}(s, t) = r_X(s, t) + r_Y(s, t).$$

These formulas are readily extended to sums of any finite number of processes. Consider in particular n second order stationary processes $X_1(t), -\infty < t < \infty, \ldots, X_n(t), -\infty < t < \infty$, whose cross-covariance functions all vanish. Then their sum

$$X(t) = X_1(t) + \cdots + X_n(t), \quad -\infty < t < \infty,$$

is a second order stationary process such that

(13) $$\mu_X = \sum_{k=1}^{n} \mu_{X_k}$$

and

(14) $$r_X(t) = \sum_{k=1}^{n} r_{X_k}(t), \quad -\infty < t < \infty.$$

Example 5. Let $Z_{11}, Z_{12}, Z_{21}, Z_{22}, \ldots, Z_{n1}, Z_{n2}$ be $2n$ independent normally distributed random variables each having mean zero and such that

$$\text{Var } Z_{k1} = \text{Var } Z_{k2} = \sigma_k^2, \quad k = 1, \ldots, n.$$

Let $\lambda_1, \ldots, \lambda_n$ be real constants and set

$$X(t) = \sum_{k=1}^{n} (Z_{k1} \cos \lambda_k t + Z_{k2} \sin \lambda_k t), \quad -\infty < t < \infty.$$

Find the mean and covariance functions of $X(t), -\infty < t < \infty$.

Set

$$X_k(t) = Z_{k1} \cos \lambda_k t + Z_{k2} \sin \lambda_k t.$$

It follows from the independence of the Z's that the cross-covariance function between $X_i(t)$ and $X_j(t)$ vanishes for $i \neq j$. Thus by using (13) and (14) together with the results of Example 1, we see that $X(t), -\infty < t < \infty$, is a second order stationary process having mean zero and covariance function

(15) $$r_X(t) = \sum_{k=1}^{n} \sigma_k^2 \cos \lambda_k t, \quad -\infty < t < \infty.$$

4.2. Gaussian processes

A stochastic process $X(t), t \in T$, is called a *Gaussian process* if every finite linear combination of the random variables $X(t), t \in T$, is normally distributed. (In this context constant random variables are regarded as normally distributed with zero variance.) Gaussian processes are also called normal processes, and normally distributed random variables are sometimes said to have a Gaussian distribution. If $X(t)$, $t \in T$, is a Gaussian process, then for each $t \in T$, $X(t)$ is normally distributed and, in particular, $EX^2(t) < \infty$. Thus a Gaussian process is necessarily a second order process. Gaussian processes have many nice theoretical properties that do not hold for second order processes in general. They are also widely used in applications, especially in engineering and in the physical sciences.

Example 6. Show that the process $X(t)$, $-\infty < t < \infty$, from Example 1 is a Gaussian process.

To verify that this is a Gaussian process, we let n be a positive integer and choose real numbers $t_1, \ldots, t_n$ and $a_1, \ldots, a_n$. Now

$$X(t) = Z_1 \cos \lambda t + Z_2 \sin \lambda t,$$

where Z_1 and Z_2 are independent and normally distributed. Thus

$a_1 X(t_1) + \cdots + a_n X(t_n)$

$= Z_1(a_1 \cos \lambda t_1 + \cdots + a_n \cos \lambda t_n) + Z_2(a_1 \sin \lambda t_1 + \cdots + a_n \sin \lambda t_n)$

is a linear combination of independent normally distributed random variables and therefore is itself normally distributed.

It is left as an exercise for the reader to show that the process in Example 5 is also a Gaussian process.

Two stochastic processes $X(t)$, $t \in T$, and $Y(t)$, $t \in T$, are said to have the same joint distribution functions if for every positive integer n and every choice of $t_1, \ldots, t_n$, all in T, the random variables

$$X(t_1), \ldots, X(t_n)$$

have the same joint distribution function as the random variables

$$Y(t_1), \ldots, Y(t_n).$$

One of the most useful properties of Gaussian processes is that if two such processes have the same mean and covariance functions, then they also have the same joint distribution functions. We omit the proof of this result. To see that the Gaussian assumption is necessary, observe that the process defined in Exercise 15 has the same mean and covariance functions as that from Example 1 with $\sigma^2 = 1$ but not the same joint distribution functions.

4.2. Gaussian processes

The mean and covariance functions can also be used to find the higher moments of a Gaussian process.

Example 7. Let $X(t), t \in T$, be a Gaussian process having zero means. Find $EX^4(t)$ in terms of the covariance function of the process.

We recall that if X is normally distributed with mean 0 and variance σ^2, then $EX^4 = 3\sigma^4$. Since $X(t)$ is normally distributed with mean 0 and variance $r_X(t, t)$, we see that

$$EX^4(t) = 3(r_X(t, t))^2.$$

Let n be a positive integer and let $X_1, \ldots, X_n$ be random variables. They are said to have a joint *normal* (or *Gaussian*) *distribution* if

$$a_1 X_1 + \cdots + a_n X_n$$

is normally distributed for every choice of the constants $a_1, \ldots, a_n$. A stochastic process $X(t), t \in T$, is a Gaussian process if and only if for every positive integer n and every choice of $t_1, \ldots, t_n$ all in T, the random variables $X(t_1), \ldots, X(t_n)$ have a joint normal distribution.

Let $X_1, \ldots, X_n$ be random variables having a joint normal distribution and a density f with respect to integration on R^n. (Such a density exists if and only if the covariance matrix of $X_1, \ldots, X_n$ has nonzero determinant.) It can be shown that f is necessarily of the form

$$(16) \quad f(x_1, \ldots, x_n) = \frac{1}{(2\pi)^{n/2}(\det \Sigma)^{1/2}} \exp\left[-\tfrac{1}{2}(\mathbf{x} - \boldsymbol{\mu})' \Sigma^{-1} (\mathbf{x} - \boldsymbol{\mu})\right],$$

where Σ is the covariance matrix

$$\Sigma = \begin{bmatrix} \operatorname{cov}(X_1, X_1) & \cdots & \operatorname{cov}(X_1, X_n) \\ \vdots & & \vdots \\ \operatorname{cov}(X_n, X_1) & \cdots & \operatorname{cov}(X_n, X_n) \end{bmatrix},$$

$\mathbf{x}$ and $\boldsymbol{\mu}$ are the vectors

$$\mathbf{x} = \begin{bmatrix} x_1 \\ \vdots \\ x_n \end{bmatrix}, \quad \boldsymbol{\mu} = \begin{bmatrix} \mu_1 \\ \vdots \\ \mu_n \end{bmatrix},$$

and $'$ denotes matrix transpose. In particular, if $n = 2$, then (16) can be written as

$$(17) \quad f(x_1, x_2) = \frac{1}{2\pi\sigma_1\sigma_2\sqrt{1-\rho^2}} \exp\left[-\frac{Q(x_1, x_2)}{2}\right],$$

where

$$Q(x_1, x_2) = \frac{1}{(1 - \rho^2)}$$
$$\times \left[\left(\frac{x_1 - \mu_1}{\sigma_1}\right)^2 - 2\rho \left(\frac{x_1 - \mu_1}{\sigma_1}\right)\left(\frac{x_2 - \mu_2}{\sigma_2}\right) + \left(\frac{x_2 - \mu_2}{\sigma_2}\right)^2 \right].$$

Here μ_1 and σ_1^2 denote the mean and variance of X_1, μ_2 and σ_2^2 denote the mean and variance of X_2, and ρ denotes the correlation between X_1 and X_2. One can also use (16) to show that the conditional expectation of X_n given $X_1, \ldots, X_{n-1}$ is a linear function of these $n - 1$ random variables, i.e., that

$$E[X_n \mid X_1 = x_1, \ldots, X_{n-1} = x_{n-1}] = a + b_1 x_1 + \cdots + b_{n-1} x_{n-1}$$

for suitable constants $a, b_1, \ldots, b_{n-1}$.

A stochastic process $X(t)$, $-\infty < t < \infty$, is said to be *strictly stationary* if for every number τ the stochastic process $Y(t)$, $-\infty < t < \infty$, defined by

$$Y(t) = X(t + \tau), \quad -\infty < t < \infty,$$

has the same joint distribution functions as the $X(t)$ process. A strictly stationary process need not have finite second moments and hence need not be a second order process. It is clear, however, that if a strictly stationary process does have finite second moments, then it is a second order stationary process. The converse is not true in general. It is left as an exercise for the reader to demonstrate by an example that a second order stationary process need not be strictly stationary.

Let $X(t)$, $-\infty < t < \infty$, be a second order stationary process which is also a Gaussian process. Then this process is necessarily strictly stationary. For if τ is any real number, then the $Y(t)$ process defined by $Y(t) = X(t + \tau)$, $-\infty < t < \infty$, is a Gaussian process having the same mean and covariance functions as the $X(t)$ process. It therefore has the same joint distribution functions as the $X(t)$ process.

Since the processes in Examples 1 and 5 are Gaussian and second order stationary, they are also strictly stationary. The second order stationary processes from Examples 2 and 4 are not Gaussian, but it can be shown that they too are strictly stationary.

4.3. The Wiener process

It has long been known from microscopic observations that particles suspended in a liquid are in a state of constant highly irregular motion. It gradually came to be realized that the cause of this motion is the bombardment of the particles by the smaller invisible molecules of the

4.3. The Wiener process

liquid. Such motion is called "Brownian motion," named after one of the first scientists to study it carefully.

Many mathematical models for this physical process have been proposed. We will now describe one such model. Let the location of a particle be described by a Cartesian coordinate system whose origin is the location of the particle at time $t = 0$. Then the three coordinates of the position of the particle vary independently, each according to a stochastic process $W(t)$, $-\infty < t < \infty$, satisfying the following properties:

(i) $W(0) = 0$.
(ii) $W(t) - W(s)$ has a normal distribution with mean 0 and variance $\sigma^2(t - s)$ for $s \leq t$.
(iii) $W(t_2) - W(t_1)$, $W(t_3) - W(t_2)$, ..., $W(t_n) - W(t_{n-1})$ are independent for $t_1 \leq t_2 \leq \cdots \leq t_n$.

Here σ^2 is some positive constant.

Property (i) follows from our choice of the coordinate system. Properties (ii) and (iii) are plausible if the motion is caused by an extremely large number of unrelated and individually negligible collisions which have no more tendency to move the particle in one direction than in the opposite direction. In particular, the central limit theorem makes it reasonable to suppose that the increments $W(t) - W(s)$ are normally distributed.

This model was initiated, in a different form, by Albert Einstein in 1905. He related the parameter σ^2 to various physical parameters including Avogadro's number. Estimation of σ^2 together with other measurements in a scientific experiment conducted shortly thereafter led to an estimate of Avogadro's number that is within 19 percent of the presently accepted value. Einstein's work and its experimental confirmation gave added evidence for the atomic basis of matter, which was still being questioned at the turn of the century.

Although the mathematical model is reasonable and fits the experimental data quite well, it has certain theoretical deficiencies that will be discussed in Section 5.3. In Chapter 6 we will discuss another mathematical model for the physical process.

A stochastic process $W(t)$, $-\infty < t < \infty$, satisfying properties (i)-(iii) is called the *Wiener process* with parameter σ^2. Mathematicians Norbert Wiener and Paul Lévy developed much of the theory, and the process is also known as the Wiener-Lévy process and as Brownian motion. The Wiener process is usually assumed to satisfy an additional property involving "continuity of the sample functions," which we will discuss in Section 5.1.2.

It follows immediately from the properties of the Wiener process that the random variables $W(t)$ all have mean 0 and that

(18) $\quad E(W(t_2) - W(t_1))(W(t_4) - W(t_3)) = 0, \quad t_1 \leq t_2 \leq t_3 \leq t_4.$

The covariance function of the process is

(19) $\quad r_W(s, t) = \begin{cases} \sigma^2 \min(|s|, |t|), & st > 0, \\ 0, & st \leq 0. \end{cases}$

The proof of (19) is virtually identical to that of Formula (10) for the covariance function of the Poisson process defined in Example 3. It is left as an exercise for the reader to show that

(20) $\quad E(W(s) - W(a))(W(t) - W(a))$
$$= \sigma^2 \min(s - a, t - a), \quad s \geq a \text{ and } t \geq a.$$

The Wiener process is a Gaussian process. In other words, if $t_1 \leq \cdots \leq t_n$ and $b_1, \ldots, b_n$ are real constants, the random variable

$$b_1 W(t_1) + \cdots + b_n W(t_n)$$

is normally distributed. In proving this result we can assume, with no loss of generality, that one of the numbers $t_1, \ldots, t_n$, say t_k, equals zero. Then each of the random variables $W(t_1), \ldots, W(t_n)$ is a linear combination of the increments $W(t_2) - W(t_1), \ldots, W(t_n) - W(t_{n-1})$. Indeed, $W(t_k) = 0$,

$$W(t_j) = (W(t_{k+1}) - W(t_k)) + \cdots + (W(t_j) - W(t_{j-1})), \quad k < j \leq n,$$

and

$$W(t_j) = (W(t_j) - W(t_{j+1})) + \cdots + (W(t_{k-1}) - W(t_k)), \quad 1 \leq j < k.$$

Thus $b_1 W(t_1) + \cdots + b_n W(t_n)$ can also be written as a linear combination of the increments $W(t_2) - W(t_1), \ldots, W(t_n) - W(t_{n-1})$. Now these increments are independent and normally distributed. Thus any linear combination of them, in particular,

$$b_1 W(t_1) + \cdots + b_n W(t_n)$$

is normally distributed.

Exercises

1 Let $X(t)$, $-\infty < t < \infty$, be a second order process. Show that it is a second order stationary process if and only if $\mu_X(t)$ is independent of t and $r_X(s, t)$ depends only on the difference between s and t.

2 Let $X(t)$, $-\infty < t < \infty$, be a second order process. Show that it is a second order stationary process if and only if $EX(s)$ and $EX(s)X(s + t)$ are both independent of s.

3 Let $X(t)$, $-\infty < t < \infty$, be a second order stationary process and set $Y(t) = X(t + 1) - X(t)$, $-\infty < t < \infty$. Show that the $Y(t)$ process is a second order stationary process having zero means and covariance function

$$r_Y(t) = 2r_X(t) - r_X(t - 1) - r_X(t + 1).$$

4 Let $X(t)$, $-\infty < t < \infty$, be a second order stationary process.
(a) Show that

$$\text{Var}(X(s + t) - X(s)) = 2(r_X(0) - r_X(t)).$$

(b) Show that for $M > 0$

$$P(|X(s + t) - X(s)| \geq M) \leq \frac{2}{M^2}(r_X(0) - r_X(t)).$$

5 Let $X(t)$, $-\infty < t < \infty$, be a Poisson process with parameter λ and set $Y(t) = X(t) - tX(1)$, $0 \leq t \leq 1$. Find the mean and covariance functions of the $Y(t)$ process.

6 Let $U_1, \ldots, U_n$ be independent random variables, each uniformly distributed on $(0, 1)$. Let $\psi(t, x)$, $0 \leq t \leq 1$ and $0 \leq x \leq 1$, be defined by

$$\psi(t, x) = \begin{cases} 1, & x \leq t, \\ 0, & x > t. \end{cases}$$

Then

$$X(t) = \frac{1}{n} \sum_{k=1}^{n} \psi(t, U_k), \quad 0 \leq t \leq 1,$$

is the *empirical distribution function* of $U_1, \ldots, U_n$. Compute the mean and covariance functions of the $X(t)$ process.

7 Let $X(t)$, $-\infty < t < \infty$, be a second order stationary process having covariance function $r_X(t)$, $-\infty < t < \infty$. Set $Y(t) = X(t + 1)$, $-\infty < t < \infty$. Find the cross-covariance function between the $X(t)$ process and the $Y(t)$ process.

8 Let R and Θ be independent random variables such that Θ is uniformly distributed on $[0, 2\pi)$ and R has the density

$$f_R(r) = \begin{cases} \dfrac{r}{\sigma^2} e^{-r^2/2\sigma^2}, & 0 < r < \infty, \\ 0, & r \leq 0, \end{cases}$$

where σ is a positive constant. It follows by using the change of variable formula involving Jacobians that $R \cos \Theta$ and $R \sin \Theta$ are independent random variables, each normally distributed with mean 0 and variance σ^2. Let λ be a positive constant and set

$$X(t) = R \cos(\lambda t + \Theta), \quad -\infty < t < \infty.$$

Show that the $X(t)$ process is a second order stationary process having mean zero and covariance function

$$r_X(t) = \sigma^2 \cos \lambda t, \quad -\infty < t < \infty.$$

9 Let $R_1, \ldots, R_n, \Theta_1, \ldots, \Theta_n$ be independent random variables such that the Θ's are uniformly distributed on $[0, 2\pi)$ and R_k has the density

$$f_{R_k}(r) = \begin{cases} \dfrac{r}{\sigma_k^2} e^{-r^2/2\sigma_k^2}, & 0 < r < \infty, \\ 0, & r \le 0, \end{cases}$$

where $\sigma_1, \ldots, \sigma_n$ are positive constants. Let $\lambda_1, \ldots, \lambda_n$ be positive constants and set

$$X(t) = \sum_{k=1}^{n} R_k \cos(\lambda_k t + \Theta_k).$$

Show that the $X(t)$ process is a second order stationary process having mean zero and covariance function

$$r_X(t) = \sum_{k=1}^{n} \sigma_k^2 \cos \lambda_k t.$$

10 Show that the $X(t)$ process in Example 5 is a Gaussian process.

11 Show that the $X(t)$ process in Exercise 9 is a Gaussian process.

12 Let $X(t)$, $-\infty < t < \infty$, be a Gaussian process and let f and g be functions from $(-\infty, \infty)$ to $(-\infty, \infty)$. Show that $Y(t) = f(t)X(g(t))$, $-\infty < t < \infty$, is a Gaussian process and find its mean and covariance functions.

13 Let $X(t)$, $-\infty < t < \infty$, be a Gaussian process having mean zero and set $Y(t) = X^2(t)$, $-\infty < t < \infty$.
   (a) Find the mean and covariance functions of the $Y(t)$ process.
   (b) Show that if the $X(t)$ process is a second order stationary process, then so is the $Y(t)$ process.

14 Let X_1 and X_2 have the joint density given by (17).
   (a) Find the conditional density of X_2 given $X_1 = x_1$.
   (b) Find the conditional expectation of X_2 given $X_1 = x_1$.

15 Let Z_1 and Z_2 be independent and identically distributed random variables taking on the values -1 and 1 each with probability $1/2$. Show that $X(t) = Z_1 \cos \lambda t + Z_2 \sin \lambda t$, $-\infty < t < \infty$, is a second order stationary process which is not strictly stationary.

Exercises

In the remaining problems $W(t)$, $-\infty < t < \infty$, is the Wiener process with parameter σ^2.

16 Verify Formula (20).

17 Find the distribution of $W(1) + \cdots + W(n)$ for a positive integer n.
Hint: Use the formulas
$$1 + 2 + \cdots + n = \frac{n(n+1)}{2}$$
and
$$1^2 + 2^2 + \cdots + n^2 = \frac{n(n+1)(2n+1)}{6}.$$

18 Set
$$X(t) = \frac{W(t+\varepsilon) - W(t)}{\varepsilon}, \qquad -\infty < t < \infty,$$
where ε is a positive constant. Show that the $X(t)$ process is a stationary Gaussian process having covariance function
$$r_X(t) = \begin{cases} \dfrac{\sigma^2}{\varepsilon}\left(1 - \dfrac{|t|}{\varepsilon}\right), & |t| < \varepsilon, \\ 0, & |t| \geq \varepsilon. \end{cases}$$

19 Set
$$X(t) = e^{-\alpha t} W(e^{2\alpha t}), \qquad -\infty < t < \infty,$$
where α is a positive constant. Show that the $X(t)$ process is a stationary Gaussian process having covariance function
$$r_X(t) = \sigma^2 e^{-\alpha|t|}, \qquad -\infty < t < \infty.$$

20 Find the mean and covariance functions of the following processes:
(a) $X(t) = (W(t))^2$, $\quad t \geq 0$;
(b) $X(t) = tW(1/t)$, $\quad t > 0$;
(c) $X(t) = c^{-1}W(c^2 t)$, $\quad t \geq 0$;
(d) $X(t) = W(t) - tW(1)$, $\quad 0 \leq t \leq 1$.

5 Continuity, Integration, and Differentiation of Second Order Processes

In this chapter we will study integration and differentiation of continuous parameter second order processes. We will see that the Wiener process is not differentiable in the ordinary sense, but leads to a new type of process called "white noise."

5.1. Continuity assumptions

In dealing with continuous parameter second order processes, it is customary to assume that their mean and covariance functions are continuous and also to make some assumptions concerning the continuity of the process itself.

5.1.1. Continuity of the mean and covariance functions.

Let $X(t)$, $t \in T$, be a continuous parameter second order process. By definition then, T is an interval having positive length. We assume in this chapter that

(i) $\mu_X(t)$, $t \in T$, is a continuous function of t

and that

(ii) $r_X(s, t)$, $s \in T$ and $t \in T$, is jointly continuous in s and t.

These assumptions are satisfied in all the examples of the previous chapter and in virtually all other examples arising in practice.

Assumptions (i) and (ii) have the interesting consequence that the process $X(t)$, $t \in T$, is *continuous in mean square*, i.e., that

(1) $$\lim_{s \to t} E(X(s) - X(t))^2 = 0, \quad t \in T.$$

5.1. Continuity assumptions

To verify (1), write

$$E(X(s) - X(t))^2 = [E(X(s) - X(t))]^2 + \text{Var}(X(s) - X(t))$$
$$= [EX(s) - EX(t)]^2 + \text{Var } X(s)$$
$$- 2 \text{ cov}(X(s), X(t)) + \text{Var } X(t)$$
$$= (\mu_X(s) - \mu_X(t))^2 + r_X(s, s) - 2r_X(s, t) + r_X(t, t).$$

It follows from (i) that

$$\lim_{s \to t} (\mu_X(s) - \mu_X(t))^2 = 0, \quad t \in T.$$

It follows from (ii) that $r_X(s, s)$ and $r_X(s, t)$ approach $r_X(t, t)$ as $s \to t$, and hence that

$$\lim_{s \to t} (r_X(s, s) - 2r_X(s, t) + r_X(t, t)) = 0.$$

Equation (1) follows immediately from these results.

Let $Y(t), t \in T$, be another continuous parameter second order process satisfying (i) and (ii). Then the cross-covariance function $r_{XY}(s, t), s \in T$ and $t \in T$, is jointly continuous in s and t. In other words,

(2) $\quad \lim_{u \to s, v \to t} \text{cov}(X(u), Y(v)) = \text{cov}(X(s), Y(t)), \quad s \in T \text{ and } t \in T.$

To verify (2) we write

$$\text{cov}(X(s), Y(t)) = EX(s)Y(t) - \mu_X(s)\mu_Y(t)$$

and

$$\text{cov}(X(u), Y(v)) = EX(u)Y(v) - \mu_X(u)\mu_Y(v).$$

The difference between these two covariances can be written as

(3) $\quad \text{cov}(X(u), Y(v)) - \text{cov}(X(s), Y(t))$
$$= E(X(u) - X(s))Y(v) + EX(s)(Y(v) - Y(t))$$
$$- (\mu_X(u) - \mu_X(s))\mu_Y(v) - \mu_X(s)(\mu_Y(v) - \mu_Y(t)).$$

It follows from (i) applied to the two processes that

(4) $\quad \lim_{u \to s, v \to t} (\mu_X(u) - \mu_X(s))\mu_Y(v) = 0$

and

(5) $\quad \lim_{v \to t} \mu_X(s)(\mu_Y(v) - \mu_Y(t)) = 0.$

By Schwarz's inequality

$$[E(X(u) - X(s))Y(v)]^2 \leq E(X(u) - X(s))^2 E(Y(v))^2$$
$$= E(X(u) - X(s))^2 (r_Y(v, v) + (\mu_Y(v))^2).$$

Thus by (1) and by (i) and (ii) applied to the $Y(t)$ process,

(6) $$\lim_{u \to s, v \to t} E(X(u) - X(s))Y(v) = 0.$$

By a similar argument

(7) $$\lim_{u \to s, v \to t} EX(s)(Y(v) - Y(t)) = 0.$$

Equation (2) follows immediately from (3)–(7).

5.1.2. Continuity of the sample functions. The random variables $X(t)$, $t \in T$, are defined on some fixed probability space Ω. Temporarily we will use the notation $X(t, \omega)$ to denote the dependence on both t and ω. For each $\omega \in \Omega$ the function $X(t, \omega)$, $t \in T$, defines a real-valued function of t, called the *sample function* of the process. Thus every $\omega \in \Omega$ is associated with a unique sample function, and we can think of a stochastic process as a random sample function.

The sample functions from Example 5 of Chapter 4 all satisfy the assumption

(iii) $X(t, \omega)$, $t \in T$, is a continuous function of t.

Assumption (iii) is certainly reasonable in models of "continuously varying" physical processes such as Brownian motion. Portions of typical sample functions of the processes in Examples 2, 3, and 4 of Chapter 4 are shown in Figure 1. The sample functions of these integer-valued processes are not continuous. They are, however, piecewise continuous and at points of discontinuity take, on their right-hand limit. Specifically, they satisfy the following three assumptions:

(iv) $X(s, \omega)$ has a finite limit as s approaches t from the left;

(v) $X(s, \omega) \to X(t, \omega)$ as s approaches t from the right;

(vi) the function $X(t, \omega)$, $t \in T$, has only a finite number of points of discontinuity on any closed, bounded subinterval of T.

In many contexts we can determine directly only that there is an event $\Omega_1 \subseteq \Omega$ such that $P(\Omega_1) = 1$ and (iv)–(vi) hold for all $\omega \in \Omega_1$. In this case we say that the sample functions satisfy (iv)–(vi) with probability one. If we now replace the probability space Ω by Ω_1, then (iv)–(vi) hold for all ω; similar remarks hold for assumption (iii). In effect we "throw out" an unwanted set of probability zero. Since this does not affect the joint distributions or the mean and covariance functions of the process, there is no reason for us to distinguish properties that hold "with probability one" from those that hold "for all $\omega \in \Omega$."

5.1. Continuity assumptions

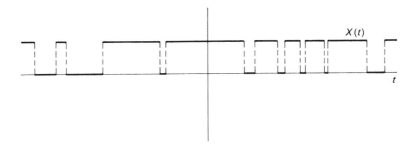

Figure 1a   Example 2 of Chapter 4

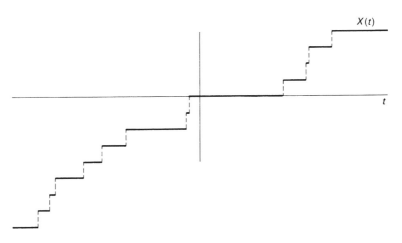

Figure 1b   Example 3 of Chapter 4

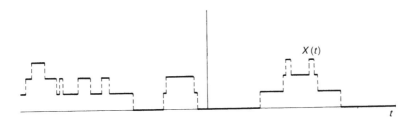

Figure 1c   Example 4 of Chapter 4

A precise discussion of sample function continuity for Gaussian processes is much more complicated than that for integer-valued processes. It has been shown that the sample functions of a continuous parameter

Gaussian process either are continuous with probability one or with probability one are so highly discontinuous as to be unlikely to arise in a practical problem. There are sufficient conditions known on a mean function and a covariance function which guarantee that there is a Gaussian process having that mean and that covariance function and having continuous sample functions. In particular, it can be shown that there is a process having the properties of a Wiener process and having continuous sample functions. Norbert Wiener first proved this result. Several proofs are now known, but they are all too difficult to be included in a book at this level.

In the future, by the Wiener process with parameter σ^2, we will mean a process satisfying properties (i)–(iii) of Section 4.3 and having continuous sample functions.

5.2. Integration

In this section we will obtain formulas for the means and covariances of random variables and processes defined in terms of a continuous parameter second order process by means of integration. For a simple illustration where integration of processes arises naturally, consider a second order stationary process $X(t)$, $-\infty < t < \infty$, having constant but unknown mean μ. Suppose we observe the process on $a \le t \le b$ and wish to estimate μ. A simple and typically very good estimate is given by

$$\hat{\mu} = \frac{1}{b-a} \int_a^b X(t)\, dt.$$

Integration of processes will be used later in defining "white noise" and in solving differential equations having "white noise" or other processes as inputs.

Let $X(t)$, $t \in T$, be a continuous parameter second order process satisfying assumptions (i), (ii), and (iv)–(vi) of Section 5.1. Let $f(t)$, $a \le t \le b$, be a piecewise continuous function, where $[a, b]$ is a closed and bounded subinterval of T. For each $\omega \in \Omega$ the function $X(t, \omega)$, $a \le t \le b$, is piecewise continuous, and hence $f(t)X(t, \omega)$, $a \le t \le b$, is also piecewise continuous. Since the ordinary integrals of calculus are well defined for piecewise continuous functions,

$$\int_a^b f(t)X(t, \omega)\, dt$$

is well defined for each $\omega \in \Omega$. By using results from measure theory, it can be shown that this integral, as a function of ω, defines a random

5.2. Integration

variable. It can also be shown that the expectation of this random variable may be found by interchanging the order of integration and expectation. In other words,

(8) $$E\left[\int_a^b f(t)X(t)\,dt\right] = \int_a^b f(t)EX(t)\,dt$$
$$= \int_a^b f(t)\mu_X(t)\,dt.$$

Here we have returned to our usual notational convention of omitting the dependence of $X(t, \omega)$ on ω.

Let $f(t)$, $a \le t \le b$, and $g(t)$, $c \le t \le d$, be two such piecewise continuous functions. In order to compute the expectation of the random variable

$$\int_a^b f(t)X(t)\,dt \int_c^d g(t)X(t)\,dt,$$

we first rewrite it as the iterated integral

$$\int_a^b f(s) \left(\int_c^d g(t)X(s)X(t)\,dt\right) ds.$$

It is again permissible to interchange the order of integration and expectation. Thus

(9) $$E\left[\int_a^b f(t)X(t)\,dt \int_c^d g(t)X(t)\,dt\right]$$
$$= \int_a^b f(s) \left(\int_c^d g(t)E[X(s)X(t)]\,dt\right) ds.$$

Using (8) and (9), we conclude that

(10) $$\operatorname{cov}\left(\int_a^b f(t)X(t)\,dt, \int_c^d g(t)X(t)\,dt\right)$$
$$= \int_a^b f(s) \left(\int_c^d g(t)r_X(s, t)\,dt\right) ds.$$

As an illustration of the use of (8) and (10), let $t_0 \in T$ and consider the process $Y(t)$, $t \in T$, defined by

$$Y(t) = \int_{t_0}^t X(s)\,ds, \quad t \in T.$$

Then by (8)

$$\mu_Y(t) = \int_{t_0}^t \mu_X(s)\,ds, \quad t \in T,$$

and by (10)

$$r_Y(s, t) = \int_{t_0}^s \left(\int_{t_0}^t r_X(u, v)\,dv\right) du, \quad s \in T \text{ and } t \in T.$$

It follows from these formulas that the $Y(t)$ process has continuous sample functions and continuous mean and covariance functions.

If $X(t)$, $t \in T$, is a Gaussian process, then

$$\int_a^b f(t)X(t)\, dt$$

has a normal distribution. In proving this result it is first necessary to approximate the integral by a finite sum such as

$$\frac{b-a}{n} \sum_{k=1}^n f\left(a + \frac{(b-a)k}{n}\right) X\left(a + \frac{(b-a)k}{n}\right).$$

This sum is normally distributed and converges to the integral as $n \to \infty$. Using this and theorems from advanced probability theory, one can show that the integral is normally distributed.

More generally, let $h(s, t)$, $s \in S$ and $a \le t \le b$, be such that for each s in the interval S the function $h(s, t)$, $a \le t \le b$, is a piecewise continuous function. Let $X(t)$, $t \in T$, be a Gaussian process and set

$$Y(s) = \int_a^b h(s, t)X(t)\, dt, \qquad s \in S.$$

Then $Y(s)$, $s \in S$, is a Gaussian process. For let $s_1, \ldots, s_n$ be in S and let $a_1, \ldots, a_n$ be real numbers. Then

$$a_1 Y(s_1) + \cdots + a_n Y(s_n) = \int_a^b \left(\sum_{k=1}^n a_k h(s_k, t) \right) X(t)\, dt$$

has a normal distribution, according to the preceding paragraph.

We will now use (9) to obtain a result that will be needed in Section 5.4. Let $W(t)$, $-\infty < t < \infty$, be the Wiener process with parameter σ^2, and let f and g be continuously differentiable functions on the closed bounded interval $[a, b]$. We will show that

(11) $$E\left[\int_a^b f'(t)(W(t) - W(a))\, dt \int_a^b g'(t)(W(t) - W(a))\, dt \right]$$
$$= \sigma^2 \int_a^b (f(t) - f(b))(g(t) - g(b))\, dt.$$

For a simple application of (11), set $a = 0$, $b = 1$, and $f(t) = g(t) = t$ for $0 \le t \le 1$. We conclude from (11) that

$$E\left(\int_0^1 W(t)\, dt \right)^2 = \sigma^2 \int_0^1 (t - 1)^2\, dt$$
$$= \frac{\sigma^2}{3}.$$

The random variable

$$\int_0^1 W(t)\, dt$$

has mean zero. It is normally distributed, since the Wiener process is a Gaussian process. Thus

$$\int_0^1 W(t)\, dt$$

is normally distributed with mean zero and variance $\sigma^2/3$.

We will now prove that (11) holds. By (9) and Formula (20) of Section 4.3 we need only show that

(12) $$\int_a^b f'(s) \left(\int_a^b g'(t) \min(s-a, t-a)\, dt \right) ds$$
$$= \int_a^b (f(t) - f(b))(g(t) - g(b))\, dt.$$

In verifying (12), we write the inner integral of the left side as

$$\int_a^s (t-a)g'(t)\, dt + (s-a) \int_s^b g'(t)\, dt.$$

Integrating the first of these two integrals by parts, we rewrite this expression as

$$(t-a)g(t)\Big|_a^s - \int_a^s g(t)\, dt + (s-a)(g(b) - g(s))$$
$$= (s-a)g(b) - \int_a^s g(t)\, dt$$
$$= \int_a^s (g(b) - g(t))\, dt.$$

Thus the left side of (12) equals

$$\int_a^b f'(s) \left(\int_a^s (g(b) - g(t))\, dt \right) ds.$$

Interchanging the order of integration, we get

$$\int_a^b (g(b) - g(t)) \left(\int_t^b f'(s)\, ds \right) dt = \int_a^b (g(b) - g(t))(f(b) - f(t))\, dt,$$

which equals the right side of (12).

5.3. Differentiation

Let $X(t)$, $t \in T$, be a continuous parameter second order process satisfying (i) and (ii). We say that this process is *differentiable* if there is a second order process $Y(t)$, $t \in T$, satisfying assumptions (i), (ii), and (iv)–(vi) of Section 5.1 and such that for $t_0 \in T$

$$X(t) - X(t_0) = \int_{t_0}^t Y(s)\, ds, \qquad t \in T.$$

The $Y(t)$ process is then called the derivative of the $X(t)$ process and is denoted by $X'(t)$, $t \in T$. Thus

(13) $$X(t) - X(t_0) = \int_{t_0}^t X'(s)\, ds, \qquad t \in T.$$

It follows from (13) that the $X(t)$ process has continuous sample functions and thus satisfies assumption (iii) of Section 5.1, and that

$$\frac{d}{dt} X(t) = X'(t)$$

holds except at the points of discontinuity of the $X'(t)$ process. It also follows from (13) that

$$\mu_X(t) - \mu_X(t_0) = \int_{t_0}^t \mu_{X'}(s)\, ds,$$

and hence that

(14) $$\mu_{X'}(t) = \frac{d}{dt} \mu_X(t), \qquad t \in T.$$

In order to find the covariance function of the $X'(t)$ process, we first consider a more general result involving cross-covariance functions. Let $X(t)$, $t \in T$, and $Y(t)$, $t \in T$, be two second order processes, and suppose that the $X(t)$ process is differentiable and that the $Y(t)$ process satisfies (i) and (ii). We will show that

(15) $$r_{YX'}(s, t) = \frac{\partial}{\partial t} r_{YX}(s, t), \qquad s \in T \text{ and } t \in T,$$

and

(16) $$r_{X'Y}(s, t) = \frac{\partial}{\partial s} r_{XY}(s, t), \qquad s \in T \text{ and } t \in T.$$

In order to verify (15) we choose $t_0 \in T$ and write

$$X(t) - X(t_0) = \int_{t_0}^t X'(u)\, du,$$

from which it follows that

$$Y(s)(X(t) - X(t_0)) = \int_{t_0}^t Y(s) X'(u)\, du.$$

5.3. Differentiation

Thus

(17) $$E[Y(s)(X(t) - X(t_0))] = \int_{t_0}^{t} E[Y(s)X'(u)]\, du.$$

Now

$$\mu_X(t) - \mu_X(t_0) = \int_{t_0}^{t} \mu_{X'}(u)\, du,$$

and hence

(18) $$\mu_Y(s)(\mu_X(t) - \mu_X(t_0)) = \int_{t_0}^{t} \mu_Y(s)\mu_{X'}(u)\, du.$$

By subtracting (18) from (17) and rewriting the resulting expression in terms of covariance functions, we conclude that

(19) $$r_{YX}(s, t) - r_{YX}(s, t_0) = \int_{t_0}^{t} r_{YX'}(s, u)\, du.$$

We saw in Section 5.1 that a cross-covariance function such as $r_{YX'}(s, u)$ is necessarily continuous. Thus for each fixed s we can differentiate (19) with respect to t, obtaining (15) as desired. Formula (16) follows from (15) by symmetry.

Let $X(t)$, $t \in T$, be a differentiable second order process. From (15) and (16) we have

(20) $$r_{XX'}(s, t) = \frac{\partial}{\partial t} r_{XX}(s, t) = \frac{\partial}{\partial t} r_X(s, t)$$

and

$$r_{X'X'}(s, t) = \frac{\partial}{\partial s} r_{XX'}(s, t),$$

which imply that the covariance function of the $X'(t)$ process is given by

(21) $$r_{X'}(s, t) = \frac{\partial^2}{\partial s\, \partial t} r_X(s, t), \qquad s \in T \text{ and } t \in T.$$

Let $X(t)$, $t \in T$, be a differentiable second order process. If the $X'(t)$ process is itself differentiable, we denote its derivative by $X''(t)$, $t \in T$. In this case we say that the $X(t)$ process is twice differentiable and that its second derivative is the $X''(t)$ process. Higher derivatives are similarly defined.

Let $X(t)$, $-\infty < t < \infty$, be a differentiable, second order stationary process. Then $\mu_{X'}(t) = 0$, $-\infty < t < \infty$, and (21) reduces to

$$r_{X'}(s, t) = \frac{\partial^2}{\partial s\, \partial t} r_X(t - s) = -r_X''(t - s).$$

Thus $X'(t)$, $-\infty < t < \infty$, is also a second order stationary process in this case and

(22) $$r_{X'}(t) = -\frac{d^2}{dt^2} r_X(t), \quad -\infty < t < \infty.$$

We see from (22) that the covariance function $r_X(t)$, $-\infty < t < \infty$, is twice continuously differentiable.

Example 1. Let $X(t)$, $-\infty < t < \infty$, be a differentiable second order stationary process. Show that $X(t)$ and $X'(t)$ are uncorrelated for all t.

From (20)

$$r_{XX'}(s, t) = \frac{\partial}{\partial t} r_X(t - s) = r'_X(t - s).$$

Thus

(23) $$r_{XX'}(t, t) = r'_X(0).$$

Differentiating the symmetry equation $r_X(t) = r_X(-t)$, we conclude that

$$r'_X(t) = -r'_X(-t),$$

and hence, by setting $t = 0$, that $r'_X(0) = 0$. It now follows from (23) that

$$\text{cov}\,(X(t), X'(t)) = r_{XX'}(t, t) = 0,$$

so, that $X(t)$ and $X'(t)$ are uncorrelated.

Let $X(t)$, $t \in T$, be a differentiable second order process. Then the random variables

$$\frac{X(t + h) - X(t)}{h}$$

converge in mean square to $X'(t)$ *as* $h \to 0$; that is,

(24) $$\lim_{h \to 0} E\left(\frac{X(t + h) - X(t)}{h} - X'(t)\right)^2 = 0, \quad t \in T.$$

In (24) it is understood that either $T = (-\infty, \infty)$ or, for t on the boundary of T, h is restricted to values such that $t + h \in T$.

To verify (24) we write

$$\frac{X(t + h) - X(t)}{h} - X'(t) = \frac{1}{h}\left(\int_t^{t+h} X'(u)\,du - hX'(t)\right)$$

$$= \frac{1}{h} \int_t^{t+h} (X'(u) - X'(t))\,du.$$

5.3. Differentiation

Thus by (9)

(25) $$E\left(\frac{X(t+h) - X(t)}{h} - X'(t)\right)^2$$
$$= \frac{1}{h^2} \int_t^{t+h} \left(\int_t^{t+h} E[(X'(u) - X'(t))(X'(v) - X'(t))] \, dv\right) du.$$

It follows from Schwarz's inequality that

(26) $$|E(X'(u) - X'(t))(X'(v) - X'(t))|$$
$$\leq \sqrt{E(X'(u) - X'(t))^2} \sqrt{E(X'(v) - X'(t))^2}.$$

According to results in Section 5.1, a second order process such as the $X'(t)$ process is continuous in mean square. This implies that for fixed t and fixed $\varepsilon > 0$ we can find a $\delta > 0$ such that

(27) $$E(X'(u) - X'(t))^2 \leq \varepsilon \quad \text{if} \quad |u - t| \leq \delta.$$

Let h be such that $-\delta \leq h \leq \delta$. Then by (26) and (27)

(28) $$|E(X'(u) - X'(t))(X'(v) - X'(t))| \leq \varepsilon$$

as u and v range from t to $t + h$. We see from (25) and (28) that

$$E\left(\frac{X(t+h) - X(t)}{h} - X'(t)\right)^2 \leq \varepsilon, \quad -\delta \leq h \leq \delta.$$

Since ε can be made arbitrarily small, this implies that (24) holds.

Let $X(t)$, $t \in T$, be a differentiable second order process which is also a Gaussian process. Then $X'(t)$, $t \in T$, is also a Gaussian process. To verify this result it is necessary to show that if $t_1, \ldots, t_n$ are in T and $a_1, \ldots, a_n$ are real constants, then

$$\sum_{i=1}^n a_i X'(t_i)$$

is normally distributed. We will indicate why this is the case, but omit a detailed proof.

It follows from (24) that

$$\lim_{h \to 0} E\left(\frac{X(t_i + h) - X(t_i)}{h} - X'(t_i)\right)^2 = 0, \quad i = 1, \ldots, n.$$

It is not difficult to conclude from this that

(29) $$\lim_{h \to 0} E\left(\sum_{i=1}^n a_i \left(\frac{X(t_i + h) - X(t_i)}{h}\right) - \sum_{i=1}^n a_i X'(t_i)\right)^2 = 0.$$

Now for each fixed h the random variable

$$\sum_{i=1}^{n} a_i \left(\frac{X(t_i + h) - X(t_i)}{h} \right)$$

is normally distributed. It can be shown that this, together with (29), implies that

$$\sum_{i=1}^{n} a_i X'(t_i)$$

is normally distributed.

It may come as a surprise to the reader that the Wiener process is not differentiable. The simplest way of seeing this is to observe from property (ii) in Section 4.3 that

$$E \left(\frac{W(t + h) - W(t)}{h} \right)^2 = \frac{\sigma^2}{|h|},$$

and hence that

(30) $$\lim_{h \to 0} E \left(\frac{W(t + h) - W(t)}{h} \right)^2 = \infty.$$

If the Wiener process were differentiable, then

$$\lim_{h \to 0} E \left(\frac{W(t + h) - W(t)}{h} - W'(t) \right)^2 = 0$$

would imply that

$$\lim_{h \to 0} E \left(\frac{W(t + h) - W(t)}{h} \right)^2 = E(W'(t))^2,$$

which would contradict (30).

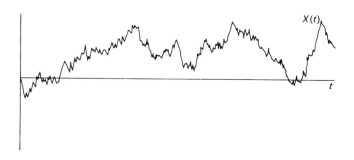

Figure 2

There are stronger senses in which the Wiener process is nondifferentiable. It has been shown that with probability one the sample function $W(t, \omega)$, $-\infty < t < \infty$, is not differentiable at even a single value of t. Thus with probability one the sample functions of the Wiener process are continuous, nowhere differentiable functions. The nondifferentiability of the Wiener process may seem especially surprising since it arose as a model for the Brownian motion of a particle. But the trajectory of a particle undergoing Brownian motion, observed under a microscope, actually does appear to be nowhere differentiable. Needless to say, it is difficult to portray a typical sample function of the Wiener process. An attempt at this has been made in Figure 2.

5.4. White noise

Let $W(t)$, $-\infty < t < \infty$, be the Wiener process with parameter σ^2. Let a and b be finite numbers and let f be a continuously differentiable function on the closed interval from a to b. Since the Wiener process is not differentiable, the integral

$$\int_a^b f(t) W'(t) \, dt \quad \text{or} \quad \int_a^b f(t) \, dW(t)$$

does not exist in the usual sense. Nevertheless, it is possible to give meaning to this integral. One way of doing so is to define the integral as

$$\lim_{\varepsilon \to 0} \int_a^b f(t) \left(\frac{W(t + \varepsilon) - W(t)}{\varepsilon} \right) dt,$$

provided the indicated limit exists. To see that this limit does indeed exist and to evaluate it explicitly, we observe that

$$\int_a^b f(t) \left(\frac{W(t + \varepsilon) - W(t)}{\varepsilon} \right) dt = \int_a^b f(t) \frac{d}{dt} \left(\frac{1}{\varepsilon} \int_t^{t+\varepsilon} W(s) \, ds \right) dt.$$

Integrating the right side of this equation by parts, we conclude that

(31) $$\int_a^b f(t) \left(\frac{W(t + \varepsilon) - W(t)}{\varepsilon} \right) dt$$

$$= \left[f(t) \frac{1}{\varepsilon} \int_t^{t+\varepsilon} W(s) \, ds \right]_a^b - \int_a^b f'(t) \left(\frac{1}{\varepsilon} \int_t^{t+\varepsilon} W(s) \, ds \right) dt.$$

Since the Wiener process has continuous sample functions, it follows that the right side of (31) converges to

$$f(t) W(t) \Big|_a^b - \int_a^b f'(t) W(t) \, dt.$$

Thus we are led to define

$$\int_a^b f(t)\,dW(t)$$

as the limit of the right side of (31) as $\varepsilon \to 0$, that is, by the formula

(32) $$\int_a^b f(t)\,dW(t) = f(b)W(b) - f(a)W(a) - \int_a^b f'(t)W(t)\,dt.$$

Note that the right side of (32) is well defined and that (32) agrees with the usual integration by parts formula.

We regard (32) as the definition of the integral appearing on the left side of (32). The derivative of the Wiener process is called "white noise." It is not a stochastic process in the usual sense. Rather $dW(t) = W'(t)\,dt$ is a "functional" that assigns values to the integral appearing on the left side of (32). Although we have given white noise a precise definition, it is not clear that it is useful for anything. We will see in Chapter 6, however, that white noise can be used to define certain stochastic differential equations, which are widely used in the physical sciences and especially in certain branches of engineering.

Since the Wiener process is a Gaussian process, it follows from (32) that

$$\int_a^b f(t)\,dW(t)$$

is normally distributed. This random variable has mean zero, as we see from (32) and the zero means of the Wiener process. To compute its variance we will show that if $a \leq b$ and g is another continuously differentiable function on $[a, b]$, then

(33) $$E\left[\int_a^b f(t)\,dW(t) \int_a^b g(t)\,dW(t)\right] = \sigma^2 \int_a^b f(t)g(t)\,dt.$$

Setting $g = f$, we see from (33) that for $a \leq b$

(34) $$\operatorname{Var}\left(\int_a^b f(t)\,dW(t)\right) = \sigma^2 \int_a^b f^2(t)\,dt.$$

We start the proof of (33) by rewriting (32) as

(35) $$\int_a^b f(t)\,dW(t)$$
$$= f(b)(W(b) - W(a)) - \int_a^b f'(t)(W(t) - W(a))\,dt.$$

5.4. White noise

By applying this formula to g as well, we conclude that

$$(36) \quad E\left[\int_a^b f(t)\,dW(t) \int_a^b g(t)\,dW(t)\right]$$

$$= E\left[\left(f(b)(W(b) - W(a)) - \int_a^b f'(t)(W(t) - W(a))\,dt\right)\right.$$

$$\left.\times \left(g(b)(W(b) - W(a)) - \int_a^b g'(t)(W(t) - W(a))\,dt\right)\right].$$

We will evaluate the right side of (36) by breaking it up into four separate terms. The product of the two integrals was computed earlier in Section 5.2. There we found that

$$(37) \quad E\left[\int_a^b f'(t)(W(t) - W(a))\,dt \int_a^b g'(t)(W(t) - W(a))\,dt\right]$$

$$= \sigma^2 \int_a^b (f(t) - f(b))(g(t) - g(b))\,dt.$$

Next we observe that by (20) of Chapter 4

$$-E\left[f(b)(W(b) - W(a)) \int_a^b g'(t)(W(t) - W(a))\,dt\right]$$

$$= -f(b) \int_a^b g'(t) E[(W(b) - W(a))(W(t) - W(a))]\,dt$$

$$= -\sigma^2 f(b) \int_a^b (t - a)g'(t)\,dt$$

$$= -\sigma^2 f(b) \left[(t - a)g(t)\Big|_a^b - \int_a^b g(t)\,dt\right]$$

$$= -\sigma^2 f(b) \left[(b - a)g(b) - \int_a^b g(t)\,dt\right].$$

Consequently

$$(38) \quad -E\left[f(b)(W(b) - W(a)) \int_a^b g'(t)(W(t) - W(a))\,dt\right]$$

$$= \sigma^2 \int_a^b f(b)(g(t) - g(b))\,dt.$$

Similarly we find that

$$(39) \quad -E\left[g(b)(W(b) - W(a))\int_a^b f'(t)(W(t) - W(a))\,dt\right]$$

$$= \sigma^2 \int_a^b g(b)(f(t) - f(b))\,dt.$$

Finally we note that

$$E[f(b)(W(b) - W(a))g(b)(W(b) - W(a))] = \sigma^2(b-a)f(b)g(b),$$

which we rewrite as

$$(40) \quad E[f(b)(W(b) - W(a))g(b)(W(b) - W(a))] = \sigma^2 \int_a^b f(b)g(b)\,dt.$$

By (36) the sum of the left sides of (37)–(40) equals the left side of (33). It is easily seen that the sum of the right sides of (37)–(40) equals the right side of (33). This proves (33).

There are two more formulas that will be needed in the next chapter:

$$(41) \quad E\left[\int_a^b f(t)\,dW(t) \int_c^d g(t)\,dW(t)\right] = 0, \quad a \le b \le c \le d,$$

and

$$(42) \quad E\left[\int_a^b f(t)\,dW(t) \int_a^c g(t)\,dW(t)\right]$$

$$= \sigma^2 \int_a^b f(t)g(t)\,dt, \quad a \le b \le c.$$

In these formulas f and g are assumed to be continuously differentiable on the indicated intervals of integration.

To verify (41) we use (35) to write

$$E\left[\int_a^b f(t)\,dW(t) \int_c^d g(t)\,dW(t)\right]$$

$$= E\left[\left(f(b)(W(b) - W(a)) - \int_a^b f'(s)(W(s) - W(a))\,ds\right) \right.$$

$$\left. \times \left(g(d)(W(d) - W(c)) - \int_c^d g'(t)(W(t) - W(c))\,dt\right)\right].$$

It follows from Formula (18) of Chapter 4 that this expectation vanishes. To verify (42) we observe first that

$$\int_a^c g(t)\,dW(t) = \int_a^b g(t)\,dW(t) + \int_b^c g(t)\,dW(t),$$

5.4. White noise

which is a direct application of the definition of these integrals given in (32). Thus

$$E\left[\int_a^b f(t)\,dW(t)\int_a^c g(t)\,dW(t)\right]$$
$$= E\left[\int_a^b f(t)\,dW(t)\left(\int_a^b g(t)\,dW(t) + \int_b^c g(t)\,dW(t)\right)\right]$$
$$= E\left[\int_a^b f(t)\,dW(t)\int_a^b g(t)\,dW(t)\right]$$
$$+ E\left[\int_a^b f(t)\,dW(t)\int_b^c g(t)\,dW(t)\right],$$

which by (33) and (41) equals the right side of (42).

Example 2. Let $X(t)$, $t \geq 0$, be defined by

$$X(t) = \int_0^t e^{\alpha(t-u)}\,dW(u), \qquad 0 \leq t < \infty,$$

where α is a real constant. Find the mean and covariance function of the $X(t)$ process.

The $X(t)$ process has zero means. For $0 \leq s \leq t$ its covariance function is found by (42) to be

$$E[X(s)X(t)] = E\left[\int_0^s e^{\alpha(s-u)}\,dW(u)\int_0^t e^{\alpha(t-u)}\,dW(u)\right]$$
$$= e^{\alpha(s+t)}E\left[\int_0^s e^{-\alpha u}\,dW(u)\int_0^t e^{-\alpha u}\,dW(u)\right]$$
$$= \sigma^2 e^{\alpha(s+t)}\int_0^s e^{-2\alpha u}\,du$$
$$= \sigma^2 e^{\alpha(s+t)}\left(\frac{1 - e^{-2\alpha s}}{2\alpha}\right)$$
$$= \frac{\sigma^2}{2\alpha}(e^{\alpha(s+t)} - e^{\alpha(t-s)}).$$

Thus by symmetry

$$r_X(s, t) = \frac{\sigma^2}{2\alpha}(e^{\alpha(s+t)} - e^{\alpha|t-s|}), \qquad s \geq 0 \text{ and } t \geq 0.$$

In particular, by setting $s = t$, we find that

$$\text{Var}(X(t)) = \frac{\sigma^2}{2\alpha}(e^{2\alpha t} - 1), \qquad t \geq 0.$$

Let f be a continuously differentiable function on $(-\infty, b]$ such that

$$\int_{-\infty}^{b} f^2(t)\, dt < \infty.$$

It can be shown that

$$\int_{-\infty}^{b} f(t)\, dW(t) = \lim_{a \to -\infty} \int_{a}^{b} f(t)\, dW(t)$$

exists and is finite with probability one, and that the random variable

$$\int_{-\infty}^{b} f(t)\, dW(t)$$

is normally distributed with mean zero and variance

$$\sigma^2 \int_{-\infty}^{b} f^2(t)\, dt.$$

Let g be a continuously differentiable function on $(-\infty, c]$ such that

$$\int_{-\infty}^{c} g^2(t)\, dt < \infty.$$

It can be shown that under these conditions (42) holds with $a = -\infty$, i.e.,

(43) $\quad E\left[\int_{-\infty}^{b} f(t)\, dW(t) \int_{-\infty}^{c} g(t)\, dW(t)\right] = \sigma^2 \int_{-\infty}^{\min(b,c)} f(t)g(t)\, dt.$

Example 3. Let $X(t)$, $-\infty < t < \infty$, be defined by

$$X(t) = \int_{-\infty}^{t} e^{\alpha(t-u)}\, dW(u), \qquad -\infty < t < \infty,$$

where α is a negative constant. Find the mean and covariance of the $X(t)$ process and show that it is a second order stationary process.

Since

$$\int_{-\infty}^{t} e^{2\alpha(t-u)}\, du = \lim_{a \to -\infty} \int_{a}^{t} e^{2\alpha(t-u)}\, du$$

$$= \lim_{a \to -\infty} \frac{1 - e^{2\alpha(t-a)}}{-2\alpha} = -\frac{1}{2\alpha}$$

is finite, we see from the remarks leading to (43) that the $X(t)$ process is well defined. It has zero means. For $s \le t$

$$r_X(s, t) = \sigma^2 \int_{-\infty}^{s} e^{\alpha(s-u)} e^{\alpha(t-u)} \, du$$

$$= \sigma^2 e^{\alpha(s+t)} \int_{-\infty}^{s} e^{-2\alpha u} \, du$$

$$= \sigma^2 e^{\alpha(s+t)} \frac{e^{-2\alpha s}}{-2\alpha}$$

$$= \frac{\sigma^2 e^{\alpha(t-s)}}{-2\alpha}.$$

In general,

$$r_X(s, t) = \frac{\sigma^2}{-2\alpha} e^{\alpha|t-s|}.$$

This shows that the $X(t)$ process is a second order stationary process having covariance function

$$r_X(t) = \frac{\sigma^2}{-2\alpha} e^{\alpha|t|}, \quad -\infty < t < \infty.$$

Summary. Under appropriate assumptions on f

(44) $$E\left[\int_{-\infty}^{\infty} f(t) \, dW(t)\right] = 0,$$

and under appropriate assumptions on f and g

(45) $$E\left[\int_{-\infty}^{\infty} f(t) \, dW(t) \int_{-\infty}^{\infty} g(t) \, dW(t)\right] = \sigma^2 \int_{-\infty}^{\infty} f(t) \, g(t) \, dt.$$

Most of these "appropriate" assumptions can be eliminated by using more sophisticated concepts. However, for (44) to hold, it is essential that

$$\int_{-\infty}^{\infty} f^2(t) \, dt < \infty,$$

and for (45) to hold, it is essential that

$$\int_{-\infty}^{\infty} f^2(t) \, dt < \infty \quad \text{and} \quad \int_{-\infty}^{\infty} g^2(t) \, dt < \infty.$$

Exercises

1 Let $X(t)$, $t \in T$, and $Y(t)$, $t \in T$, be continuous parameter second order processes whose mean and covariance functions satisfy conditions (i) and (ii) of Section 5.1. Show that the mean and covariance functions of $Z(t) = X(t) + Y(t)$, $t \in T$, also satisfy (i) and (ii).

2 Find the correlation between $W(t)$ and

$$\int_0^1 W(s)\, ds$$

for $0 \le t \le 1$.

3 Find the mean and variance of

$$\int_0^1 W^2(t)\, dt.$$

4 Set

$$X(t) = \int_0^t W(s)\, ds, \qquad t \ge 0.$$

Find the mean and covariance function of the $X(t)$ process.

5 Set

$$X(t) = \int_t^{t+1} (W(s) - W(t))\, ds, \qquad -\infty < t < \infty.$$

Show that this is a second order stationary process having mean zero and find $r_X(t)$, $-\infty < t < \infty$.

6 Let $X(t)$, $-\infty < t < \infty$, be a second order stationary process satisfying assumptions (ii) and (iv)–(vi).

(a) Show that

$$\operatorname{Var}\left(\frac{1}{T}\int_0^T X(t)\, dt\right) = \frac{2}{T}\int_0^T r_X(t)\left(1 - \frac{t}{T}\right) dt, \qquad T > 0.$$

(b) Show that

$$\operatorname{Var}\left(\frac{1}{T}\int_0^T X(t)\, dt\right) \le r_X(0), \qquad T > 0.$$

(c) Show that

$$\operatorname{Var}\left(\frac{1}{T}\int_0^T X(t)\, dt\right) \le \frac{2}{T}\int_0^T |r_X(t)|\, dt,$$

and hence that if $\lim_{t \to \infty} r_X(t) = 0$, then

$$\lim_{T \to \infty} \operatorname{Var}\left(\frac{1}{T}\int_0^T X(t)\, dt\right) = 0.$$

7 Let $X(t)$, $-\infty < t < \infty$, be as in Exercise 6 and suppose that

$$\int_0^\infty |r_X(t)|\, dt < \infty.$$

Use the result of Exercise 6(a) to show that

$$\lim_{T \to \infty} T \, \text{Var}\left(\frac{1}{T}\int_0^T X(t)\, dt\right) = 2\int_0^\infty r_X(t)\, dt = \int_{-\infty}^\infty r_X(t)\, dt.$$

Hint: Observe that for $0 < \delta < 1$

$$\left|\int_0^T r_X(t)\,\frac{t}{T}\, dt\right| \leq \delta \int_0^\infty |r_X(t)|\, dt + \int_{\delta T}^\infty |r_X(t)|\, dt.$$

8 Let $X(t)$, $-\infty < t < \infty$, be a stationary Gaussian process satisfying properties (i)–(iii) and such that $\lim_{t \to \infty} r_X(t) = 0$. Show that if $EX(t) \equiv 0$, then

$$\lim_{T \to \infty} E\left(\frac{1}{T}\int_0^T X^2(t)\, dt - \text{Var } X(0)\right)^2 = 0.$$

Hint: Use Exercise 6 and Exercise 13 of Chapter 4.

9 Let $X(t)$, $-\infty < t < \infty$, be a second order stationary process satisfying assumptions (iv)–(vi) and having constant but unknown mean μ and covariance function

$$r_X(t) = \alpha e^{-\beta|t|}, \qquad -\infty < t < \infty,$$

where α and β are positive constants. For $T > 0$ set

$$\bar{X} = \frac{1}{T}\int_0^T X(t)\, dt.$$

(a) Show that $\bar{X}$ is an unbiased estimator of μ (i.e., $E\bar{X} = \mu$) and that

$$\text{Var}(\bar{X}) = 2\alpha\left[\frac{1}{\beta T} - \frac{1}{\beta^2 T^2}(1 - e^{-\beta T})\right].$$

(b) Set

$$\hat{\mu} = \frac{X(0) + X(T) + \beta \int_0^T X(t)\, dt}{2 + \beta T}.$$

Show that $\hat{\mu}$ is an unbiased estimator of μ and that

$$\text{Var}(\hat{\mu}) = \frac{2\alpha}{2 + \beta T}.$$

It can be shown that $\hat{\mu}$ has minimum variance among all "linear" unbiased estimators of μ based on $X(t)$, $0 \leq t \leq T$. Since Var $(\bar{X})$ is almost as small as Var $(\hat{\mu})$, the sample mean $\bar{X}$ is a very "efficient" linear estimator of μ.

(c) Show that
$$\lim_{T \to \infty} \frac{\operatorname{Var}(\hat{\mu})}{\operatorname{Var}(\bar{X})} = 1.$$
This says that $\bar{X}$ is an "asymptotically efficient" estimator of μ.

10 Let $X(t)$, $-\infty < t < \infty$, be an n-times differentiable second order process and let $Y(t)$, $-\infty < t < \infty$, be an m-times differentiable second order process. Show that
$$r_{X^{(n)}Y^{(m)}}(s, t) = \frac{\partial^{n+m}}{\partial s^n \partial t^m} r_{XY}(s, t).$$

11 Let $X(t)$, $-\infty < t < \infty$, be an n-times differentiable second order stationary process. Show that $X^{(n)}(t)$, $-\infty < t < \infty$, is a second order stationary process and that
$$r_{X^{(n)}}(t) = (-1)^n r_X^{(2n)}(t).$$

12 Let $X(t)$, $-\infty < t < \infty$, be a twice differentiable second order stationary process. In terms of $r_X(t)$, $-\infty < t < \infty$, find:
(a) $r_{XX''}(s, t)$,
(b) $r_{X'X''}(s, t)$,
(c) $r_{X''}(s, t)$.

13 Let $X(t)$, $-\infty < t < \infty$, be as in Exercise 12 and set $Y(t) = X''(t) + X(t)$, $-\infty < t < \infty$. Show that the $Y(t)$ process is a second order stationary process, and find its mean and covariance function in terms of those of the $X(t)$ process.

14 Find
$$\int_a^b c \, dW(t)$$
explicitly in terms of the Wiener process.

15 Find the mean and variance of
$$X = \int_0^1 t \, dW(t) \quad \text{and} \quad Y = \int_0^1 t^2 \, dW(t),$$
and find the correlation between these two random variables.

16 Find the covariance function of the $X(t)$ process in each of the following cases:

(a) $\quad X(t) = \int_0^t s \, dW(s), \quad\quad t \geq 0;$

(b) $\quad X(t) = \int_0^1 \cos ts \, dW(s), \quad -\infty < t < \infty;$

(c) $\quad X(t) = \int_{t-1}^t (t-s) \, dW(s), \quad -\infty < t < \infty.$

17 Let $X(t)$, $0 \leq t < \infty$, be the process defined in Example 2 and set

$$Y(t) = \int_0^t X(s)\, ds, \qquad t \geq 0.$$

(a) Show that

$$Y(t) = \int_0^t \left(\frac{e^{\alpha(t-u)} - 1}{\alpha}\right) dW(u), \qquad t \geq 0.$$

(b) Find Var $Y(t)$.

6 Stochastic Differential Equations, Estimation Theory, and Spectral Distributions

Recall that we introduced the Wiener process as a mathematical model for the motion of a particle subject to molecular bombardment. By using stochastic differential equations we can find other models for this physical process.

Let $X(t)$ represent the position at time t of a particle which moves along a straight line (alternatively, we could let $X(t)$ represent one coordinate of the position in space). Then $X'(t)$ and $X''(t)$ represent the velocity and acceleration of the particle at time t. Let m denote the mass of the particle and let $F(t)$ denote the force acting on the particle at time t. By Newton's law

(1) $$F(t) = mX''(t).$$

We will consider three types of forces:

(i) a frictional force $-fX'(t)$, due to the viscosity of the medium, proportional to the velocity and having opposite direction;
(ii) a restoring force $-kX(t)$, as in a pendulum or spring, proportional to the distance from the origin and directed toward the origin;
(iii) an external force $\xi(t)$, independent of the motion of the particle.

In short we consider a total force of the form

(2) $$F(t) = -fX'(t) - kX(t) + \xi(t),$$

where f and k are nonnegative constants. We combine (1) and (2) to obtain the differential equation

(3) $$mX''(t) + fX'(t) + kX(t) = \xi(t).$$

Suppose that the external force is due to some random effect. Then we can think of $\xi(t)$ as a stochastic process. In this case $X(t)$ is also a stochastic process and (3) is a *stochastic differential equation* relating these two processes. If the external force is due to molecular bombardment, then physical reasoning leads to the conclusion that this external force is of the form of white noise with a suitable

Introduction

parameter σ^2. In this case $X(t)$ is a stochastic process satisfying the stochastic differential equation

(4) $$mX''(t) + fX'(t) + kX(t) = W'(t),$$

where $W'(t)$ is white noise with parameter σ^2. In Sections 6.1 and 6.2, when we discuss differential equations such as (4) which involve white noise, we will define precisely what is meant by a solution to such a differential equation.

There are areas other than those related directly to molecular bombardment of particles where stochastic differential equations involving white noise arise. Consider, for example, the simple electrical circuit shown in Figure 1 consisting of a

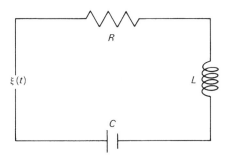

Figure 1

resistance R, an inductance L, a capacitance C, and a driving electromotive force $\xi(t)$ in series. Let $X(t)$ denote the voltage drop across the capacitor at time t. By Kirchhoff's second law $X(t)$ satisfies the differential equation

$$LCX''(t) + RCX'(t) + X(t) = \xi(t).$$

Even in the absence of a driving electromotive force there will still be a small voltage source known as "thermal noise" due to the thermal agitation of the electrons in the resistor. Physical reasoning leads to the conclusion that this thermal noise is also of the form of white noise. In this case the voltage drop satisfies the stochastic differential equation

$$LCX''(t) + RCX'(t) + X(t) = W'(t).$$

One can obtain higher order differential equations by considering more complicated electrical or mechanical systems. We will consider an nth order stochastic differential equation

(5) $$a_0 X^{(n)}(t) + a_1 X^{(n-1)}(t) + \cdots + a_n X(t) = W'(t),$$

where $a_0, a_1, \ldots, a_n$ are real constants with $a_0 \neq 0$ and $W'(t)$ is white noise with parameter σ^2.

In Section 6.1 we will consider (5) in detail for $n = 1$. There we will be able to illustrate the techniques for handling (5) in the simplest possible setting.

In Section 6.2 we will describe the corresponding results for general n, giving the full solution to (5) for $n = 2$. We will also describe what happens when the right side of (5) is replaced by a second order stationary process.

In Section 6.3 we will discuss some elementary principles of estimation theory. We will illustrate these principles by using them to predict in an optimal manner future values of solutions to stochastic differential equations.

In Section 6.4 we will describe the use of Fourier transforms in computing covariance functions of second order stationary processes. As an application of these techniques we will compute the Fourier transform of the covariance function of a second order stationary process arising as a solution to (5).

6.1. First order differential equations

In this section we will consider processes which satisfy the first order stochastic differential equation

(6) $$a_0 X'(t) + a_1 X(t) = W'(t),$$

where a_0 and a_1 are real constants with $a_0 \neq 0$ and $W'(t)$ is white noise with parameter σ^2.

For an example of such a process, let $X(t)$ be the position process governed by

$$mX''(t) + fX'(t) + kX(t) = W'(t).$$

If there is no restoring force, $k = 0$ and this equation becomes

$$mX''(t) + fX'(t) = W'(t).$$

Let $V(t) = X'(t)$ denote the velocity of the particle at time t. Since $V'(t) = X''(t)$, we see that the velocity process satisfies the differential equation

$$mV'(t) + fV(t) = W'(t),$$

which is of the same form as (6). Integrating the velocity process recovers the original position process. One can also find an example of a process satisfying a first order stochastic differential equation by considering the voltage process when there is no inductance in the network.

In trying to find solutions to (6) we first observe that it is not really well defined, since white noise does not exist as a stochastic process having

6.1. First order differential equations

sample functions in the usual sense. Ignoring this difficulty for the moment, we "formally" integrate both sides of (6) from t_0 to t and obtain

$$(7) \quad a_0(X(t) - X(t_0)) + a_1 \int_{t_0}^{t} X(s)\, ds = W(t) - W(t_0).$$

Equation (7) *is* well defined, since for any point $\omega \in \Omega$, the Wiener process sample function $W(t) = W(t, \omega)$ is a well defined continuous function. By a solution to (6) on an interval containing the point t_0, we mean a stochastic process $X(t)$ defined on that interval having continuous sample functions and satisfying (7).

In order to solve (7) we proceed through a series of reversible steps. We first rewrite this equation as

$$X(t) + \frac{a_1}{a_0} \int_{t_0}^{t} X(s)\, ds = X(t_0) - \frac{W(t_0)}{a_0} + \frac{W(t)}{a_0}.$$

Multiplying both sides of this equation by $e^{-\alpha t}$, where

$$\alpha = -\frac{a_1}{a_0},$$

we find that

$$e^{-\alpha t} X(t) - \alpha e^{-\alpha t} \int_{t_0}^{t} X(s)\, ds = \left(X(t_0) - \frac{W(t_0)}{a_0}\right) e^{-\alpha t} + \frac{e^{-\alpha t}}{a_0} W(t),$$

which we rewrite as

$$\frac{d}{dt}\left(e^{-\alpha t} \int_{t_0}^{t} X(s)\, ds\right) = \left(X(t_0) - \frac{W(t_0)}{a_0}\right) e^{-\alpha t} + \frac{e^{-\alpha t}}{a_0} W(t).$$

Integrating both sides of this equation from t_0 to t, we conclude that

$$e^{-\alpha t} \int_{t_0}^{t} X(s)\, ds = \left(X(t_0) - \frac{W(t_0)}{a_0}\right)\left(\frac{e^{-\alpha t_0} - e^{-\alpha t}}{\alpha}\right) + \int_{t_0}^{t} \frac{e^{-\alpha s}}{a_0} W(s)\, ds,$$

or equivalently,

$$\int_{t_0}^{t} X(s)\, ds = \left(X(t_0) - \frac{W(t_0)}{a_0}\right)\left(\frac{e^{\alpha(t-t_0)} - 1}{\alpha}\right) + e^{\alpha t} \int_{t_0}^{t} \frac{e^{-\alpha s}}{a_0} W(s)\, ds.$$

By differentiation we see that

$$(8) \quad X(t) = \left(X(t_0) - \frac{W(t_0)}{a_0}\right) e^{\alpha(t-t_0)} + \frac{W(t)}{a_0} + \frac{\alpha}{a_0} \int_{t_0}^{t} e^{\alpha(t-s)} W(s)\, ds.$$

Conversely, since these steps are all reversible, we see that for any choice of $X(t_0)$, the right side of (8) defines a solution to (7). Thus (8) represents the general form of the solution to (7).

By using (32) of Chapter 5, we can rewrite (8) as

(9) $$X(t) = X(t_0)e^{\alpha(t-t_0)} + \frac{1}{a_0}\int_{t_0}^t e^{\alpha(t-s)}\,dW(s).$$

Let C be any random variable. The process defined by

$$X(t) = Ce^{\alpha(t-t_0)} + \frac{1}{a_0}\int_{t_0}^t e^{\alpha(t-s)}\,dW(s)$$

is such that $X(t_0) = C$, and hence (9) holds. It is the unique stochastic process satisfying (7) and the initial condition $X(t_0) = C$. The randomness of the solution to (6) can be thought of as being caused by both the white noise term in the differential equation (6) and the randomness of the initial condition.

In many applications the initial value $X(t_0)$ is just some constant x_0 independent of $\omega \in \Omega$. In this case (9) becomes

(10) $$X(t) = x_0 e^{\alpha(t-t_0)} + \frac{1}{a_0}\int_{t_0}^t e^{\alpha(t-s)}\,dW(s).$$

This process is a Gaussian process and its mean and covariance functions are readily computed. Suppose for simplicity that $X(t)$, $t \geq 0$, is the solution to (6) on $[0, \infty)$ satisfying the initial condition $X(0) = x_0$. Then

$$X(t) = x_0 e^{\alpha t} + \frac{1}{a_0}\int_0^t e^{\alpha(t-s)}\,dW(s), \qquad t \geq 0.$$

Since integrals with respect to white noise have mean zero,

(11) $$\mu_X(t) = x_0 e^{\alpha t}, \qquad t \geq 0.$$

From Example 2 of Chapter 5 and the formula $-2\alpha a_0^2 = 2a_0 a_1$, we see that

(12) $$r_X(s, t) = \frac{\sigma^2}{2a_0 a_1}(e^{\alpha|t-s|} - e^{\alpha(s+t)}), \qquad s \geq 0 \text{ and } t \geq 0.$$

In particular,

$$\text{Var } X(t) = \frac{\sigma^2}{2a_0 a_1}(1 - e^{2\alpha t}), \qquad t \geq 0.$$

We assume throughout the remainder of this section that $\alpha = -a_1/a_0$ is negative. Then

(13) $$X_0(t) = \frac{1}{a_0}\int_{-\infty}^t e^{\alpha(t-s)}\,dW(s)$$

6.1. First order differential equations

is well defined as we saw in our discussion of Example 3 of Chapter 5. Also for $-\infty < t < \infty$

$$X_0(t) = \frac{1}{a_0} \int_{-\infty}^{0} e^{\alpha(t-s)} \, dW(s) + \frac{1}{a_0} \int_{0}^{t} e^{\alpha(t-s)} \, dW(s)$$

$$= X_0(0)e^{\alpha t} + \frac{1}{a_0} \int_{0}^{t} e^{\alpha(t-s)} \, dW(s),$$

which agrees with (9) for $t_0 = 0$. Thus $X_0(t)$, $-\infty < t < \infty$, satisfies (6) on $(-\infty, \infty)$. Our goal is to demonstrate that $X_0(t)$ is the only second order stationary process to do so. We see from Example 3 of Chapter 5 that $X_0(t)$ is a second order stationary process having zero means and covariance function

(14) $$r_{X_0}(t) = -\frac{\sigma^2}{2\alpha a_0^2} e^{\alpha|t|}$$

$$= \frac{\sigma^2}{2a_0 a_1} e^{\alpha|t|}, \quad -\infty < t < \infty.$$

It is not difficult to show that the $X_0(t)$ process is a Gaussian process.

Let $X(t)$, $-\infty < t < \infty$, be any solution to (6) on $(-\infty, \infty)$. Then for $-\infty < t < \infty$

$$X(t) = X(0)e^{\alpha t} + \frac{1}{a_0} \int_{0}^{t} e^{\alpha(t-s)} \, dW(s)$$

and

$$X_0(t) = X_0(0)e^{\alpha t} + \frac{1}{a_0} \int_{0}^{t} e^{\alpha(t-s)} \, dW(s).$$

By subtracting the second equation from the first, we conclude that

(15) $$X(t) = (X(0) - X_0(0))e^{\alpha t} + X_0(t).$$

In other words, by letting C denote the random variable $X(0) - X_0(0)$, we find that

(16) $$X(t) = Ce^{\alpha t} + X_0(t).$$

Conversely, if C is any random variable, then (16) represents a solution to (6). For (15) follows from (16) and the remainder of the above steps are reversible. We see, therefore, that (16) represents the general solution to (6), where C denotes an arbitrary random variable.

We will show next that the $X(t)$ process given by (16) is a second order stationary process if and only if $C = 0$ with probability one. Since exceptional sets of probability zero are of no concern, we can restate this result by saying that the unique second order stationary process which satisfies (6) on $(-\infty, \infty)$ is given by (13), which has zero means and covariance function given by (14).

To verify these results, let C be a random variable such that the $X(t)$ process given by (16) is a second order stationary process. Solving (16) for C, we find that

$$C = e^{-\alpha t}(X(t) - X_0(t)).$$

It follows (see Exercise 1) that

$$EC^2 \leq 2e^{-2\alpha t}(E(X(t))^2 + E(X_0(t))^2).$$

Now $E(X(t))^2 = E(X(0))^2$ and $E(X_0(t))^2 = E(X_0(0))^2$, since $X(t)$, $-\infty < t < \infty$, and $X_0(t)$, $-\infty < t < \infty$, are second order stationary processes. Thus

$$EC^2 \leq 2e^{-2\alpha t}(E(X(0))^2 + E(X_0(0))^2).$$

Letting $t \to -\infty$ in this inequality, and recalling that $\alpha < 0$, we conclude that $EC^2 = 0$. This implies that $C = 0$ with probability one, as desired.

Let $X(t)$ be a second order process satisfying (6) on $[0, \infty)$. Then (16) holds on $[0, \infty)$, where C is a random variable having finite second moment. Thus for $t \geq 0$

$$E(X(t) - X_0(t))^2 = E(Ce^{\alpha t})^2 = e^{2\alpha t}EC^2,$$

and hence

(17) $$\lim_{t \to +\infty} E(X(t) - X_0(t))^2 = 0.$$

Since

$$E(X(t) - X_0(t))^2 = (EX(t) - EX_0(t))^2 + \text{Var}(X(t) - X_0(t))$$
$$= (EX(t))^2 + \text{Var}(X(t) - X_0(t)) \geq (EX(t))^2,$$

we see from (17) that

(18) $$\lim_{t \to +\infty} \mu_X(t) = 0.$$

It follows from (17) and Schwarz's inequality (see the proof of Equation (2) of Chapter 5) that

(19) $$\lim_{s,t \to +\infty} (r_X(s, t) - r_{X_0}(s, t)) = 0.$$

We summarize (17)–(19): any second order process $X(t)$ that satisfies (6) on $[0, \infty)$ is asymptotically equal to the second order stationary solution $X_0(t)$ of (6) on $(-\infty, \infty)$, which has zero means and covariance function given by (14).

Example 1. Let $X(t)$, $0 \leq t < \infty$, be the solution to (6) on $[0, \infty)$ satisfying the initial condition $X(0) = x_0$, where x_0 is some real constant. From (11), (12), and (14), we see directly that

$$\lim_{t \to +\infty} \mu_X(t) = \lim_{t \to +\infty} x_0 e^{\alpha t} = 0$$

and that

$$\lim_{s,t \to +\infty} (r_X(s, t) - r_{X_0}(s, t)) = \lim_{s,t \to +\infty} \frac{-\sigma^2}{2a_0 a_1} e^{\alpha(s+t)} = 0.$$

6.2. Differential equations of order n

In this section we will describe the extensions of the results of Section 6.1 to solutions of nth order stochastic differential equations. Before doing so, however, we will briefly review the deterministic theory in a form convenient for our purposes.

Consider the homogeneous differential equation

$$(20) \qquad a_0 x^{(n)}(t) + a_1 x^{(n-1)}(t) + \cdots + a_n x(t) = 0,$$

where $a_0, a_1, \ldots, a_n$ are real constants with $a_0 \neq 0$. By a solution to (20) on an interval, we mean a function $\phi(t)$ which is n times differentiable and such that

$$a_0 \phi^{(n)}(t) + a_1 \phi^{(n-1)}(t) + \cdots + a_n \phi(t) = 0$$

on that interval.

For each j, $1 \leq j \leq n$, there is a solution ϕ_j to the homogeneous differential equation on $(-\infty, \infty)$ such that

$$\phi_j^{(k)}(0) = \begin{cases} 1, & k = j - 1, \\ 0, & 0 \leq k \leq n - 1 \end{cases} \quad \text{and} \quad k \neq j - 1.$$

These functions are real-valued. If $n = 1$, then $\phi_1(t) = e^{\alpha t}$, where $\alpha = -a_1/a_0$. In Section 6.2.1 we will find formulas for ϕ_1 and ϕ_2 when $n = 2$.

For any choice of the n numbers $c_1, \ldots, c_n$, the function

$$\phi(t) = c_1 \phi_1(t - t_0) + \cdots + c_n \phi_n(t - t_0)$$

is the unique solution to (20) satisfying the initial conditions

$$\phi(t_0) = c_1, \quad \phi'(t_0) = c_2, \ldots, \phi^{(n-1)}(t_0) = c_n.$$

We can write this solution in the form

$$(21) \qquad \phi(t) = \phi(t_0) \phi_1(t - t_0) + \cdots + \phi^{(n-1)}(t_0) \phi_n(t - t_0).$$

The polynomial

$$p(r) = a_0 r^n + a_1 r^{n-1} + \cdots + a_n$$

is called the *characteristic polynomial* of the left side of (20). By the fundamental theorem of algebra, it can be factored as

$$p(r) = a_0(r - r_1) \cdots (r - r_n),$$

where $r_1, \ldots, r_n$ are roots of the equation $p(r) = 0$. These roots are not necessarily distinct and may be complex-valued. If the roots are distinct, then

$$e^{r_1 t}, e^{r_2 t}, \ldots, e^{r_n t}$$

are solutions to (20), and any solution to (20) can be written as a linear combination of these solutions (i.e., these solutions form a basis for the space of all solutions to (20)). If root r_i is repeated n_i times in the factorization of the characteristic polynomial, then

$$e^{r_i t}, t e^{r_i t}, \ldots, t^{n_i - 1} e^{r_i t}$$

are all solutions to (20). As i varies, we obtain $\sum_i n_i = n$ solutions in this way, which again form a basis for the space of all solutions to (20).

The left side of (20) is *stable* if every solution to (20) vanishes at ∞. The specific form of the solutions to (20) described in the previous paragraph shows that the left side of (20) is stable if and only if the roots of the characteristic polynomial all have negative real parts.

Consider next the nonhomogeneous differential equation

(22) $$a_0 x^{(n)}(t) + a_1 x^{(n-1)}(t) + \cdots + a_n x(t) = y(t)$$

for a continuous function $y(t)$. To find the general solution to (22) on an interval, we need only find one solution to (22) and add the general solution to the corresponding homogeneous differential equation.

One method of finding a specific solution to (22) involves the *impulse response function* $h(t)$, $t \geq 0$, defined as that solution to the homogeneous differential equation (20) satisfying the initial conditions

$$h(0) = \cdots = h^{(n-2)}(0) = 0 \quad \text{and} \quad h^{(n-1)}(0) = \frac{1}{a_0}.$$

It is convenient to define $h(t)$ for all t by setting $h(t) = 0$, $t < 0$. It follows from (21) that

$$h(t) = \begin{cases} \phi_n(t)/a_0, & t \geq 0, \\ 0, & t < 0. \end{cases}$$

The function

$$x(t) = \int_{t_0}^t h(t - s) y(s) \, ds$$

6.2. Differential equations of order n

is easily shown to be the solution to (22) on an interval containing t_0 as its left endpoint and satisfying the initial conditions

$$x(t_0) = \cdots = x^{(n-1)}(t_0) = 0.$$

Suppose now that the left side of (22) is stable. Then $h(t) \to 0$ "exponentially fast" as $t \to \infty$ and, in particular,

(23) $$\int_{-\infty}^{\infty} |h(t)|\, dt < \infty \quad \text{and} \quad \int_{-\infty}^{\infty} h^2(t)\, dt < \infty.$$

If $y(t)$, $-\infty < t < \infty$, is continuous and does not grow too fast as $t \to -\infty$, e.g., if

$$\int_{-\infty}^{0} e^{ct}|y(t)|\, dt < \infty \quad \text{for all} \quad c > 0,$$

then

$$x(t) = \int_{-\infty}^{t} h(t-s)y(s)\, ds$$

$$= \int_{-\infty}^{\infty} h(t-s)y(s)\, ds$$

defines a solution to (22) on $(-\infty, \infty)$. (The reason $h(t)$ is called the "impulse response function" is that if $y(t)$, $-\infty < t < \infty$, is a "unit impulse at time 0," then the solution to (22) is

$$x(t) = \int_{-\infty}^{\infty} h(t-s)y(s)\, ds = h(t),$$

so that $h(t)$ is the response at time t to a unit impulse at time 0.)

With this background we are now ready to discuss the nth order stochastic differential equation

(24) $$a_0 X^{(n)}(t) + a_1 X^{(n-1)}(t) + \cdots + a_n X(t) = W'(t),$$

where $W'(t)$ is white noise with parameter σ^2. This equation is not well defined in its original form. We say that the stochastic process $X(t)$ is a solution to (24) on an interval containing the point t_0 if it is $n - 1$ times differentiable on that interval and satisfies the integrated form of (24), namely,

(25) $$a_0(X^{(n-1)}(t) - X^{(n-1)}(t_0)) + \cdots + a_{n-1}(X(t) - X(t_0))$$

$$+ a_n \int_{t_0}^{t} X(s)\, ds = W(t) - W(t_0)$$

on that interval.

Theorem 1 The process $X(t)$, $t \geq t_0$, defined by

$$X(t) = \int_{t_0}^{t} h(t - s) \, dW(s), \quad t \geq t_0,$$

is a solution to (24) on $[t_0, \infty)$ satisfying the initial conditions

$$X(t_0) = \cdots = X^{(n-1)}(t_0) = 0.$$

Proof. This result is just what one would expect knowing the deterministic theory. In our proof we assume for simplicity that $t_0 = 0$. If $n = 1$, then $h(t) = e^{\alpha t}/a_0$ and Theorem 1 agrees with the results found in Section 6.1. We assume from now on that $n \geq 2$. Then

$$X(t) = \int_0^t h(t - s) \, dW(s),$$

which by Equation (32) of Chapter 5 can be rewritten as

$$X(t) = h(0)W(t) + \int_0^t h'(t - s)W(s) \, ds.$$

Since $h(0) = 0$, we see that

(26) $$X(t) = \int_0^t h'(t - s)W(s) \, ds.$$

It follows from (26) that

$$\int_0^t X(s) \, ds = \int_0^t \left(\int_0^s h'(s - u)W(u) \, du \right) ds$$

$$= \int_0^t W(u) \left(\int_u^t h'(s - u) \, ds \right) du$$

$$= \int_0^t W(u)(h(t - u) - h(0)) \, du.$$

We replace the dummy variable u by s in the last integral, note again that $h(0) = 0$, and obtain

(27) $$\int_0^t X(s) \, ds = \int_0^t h(t - s)W(s) \, ds.$$

In order to find $X'(t)$ from (26), we will use the calculus formula

(28) $$\frac{d}{dt} \int_{t_0}^t f(s, t) \, ds = f(t, t) + \int_{t_0}^t \frac{\partial}{\partial t} f(s, t) \, ds,$$

which is a consequence of the chain rule. It follows from (26) and (28) that

$$X'(t) = h'(0)W(t) + \int_0^t h''(t - s)W(s) \, ds.$$

6.2. Differential equations of order n

If $n \geq 2$, then $h'(0) = 0$, and hence

$$X'(t) = \int_0^t h''(t - s)W(s)\, ds.$$

By repeated differentiation we conclude that

(29) $\quad X^{(j)}(t) = \int_0^t h^{(j+1)}(t - s)W(s)\, ds, \qquad 0 \leq j \leq n - 1.$

Since $h^{(n-1)}(0) = 1/a_0$, we find by differentiating (29) with $j = n - 2$ that

(30) $\quad X^{(n-1)}(t) = \dfrac{W(t)}{a_0} + \int_0^t h^{(n)}(t - s)W(s)\, ds.$

From (29) and (30), we see that

(31) $\quad X(0) = X'(0) = \cdots = X^{(n-1)}(0) = 0.$

It follows from (27), (29), and (30) that

$$a_0 X^{(n-1)}(t) + \cdots + a_{n-1} X(t) + a_n \int_0^t X(s)\, ds$$

$$= W(t) + \int_0^t (a_0 h^{(n)}(t - s) + \cdots + a_n h(t - s))W(s)\, ds.$$

Since $h(t)$ satisfies the homogeneous differential equation (20), the last integral vanishes, and hence

(32) $\quad a_0 X^{(n-1)}(t) + \cdots + a_{n-1} X(t) + a_n \int_0^t X(s)\, ds = W(t).$

We see from (31) and (32) that (25) holds with $t_0 = 0$. This completes the proof of the theorem. ∎

The general solution to (24) on $[t_0, \infty)$ is given by

(33) $\quad X(t) = X(t_0)\phi_1(t - t_0) + \cdots + X^{(n-1)}(t_0)\phi_n(t - t_0)$

$$+ \int_{t_0}^t h(t - s)\, dW(s), \qquad t \geq t_0.$$

In more detail, let $C_1, \ldots, C_n$ be any n random variables. Then the process $X(t)$, $t \geq t_0$, defined by

(34) $\quad X(t) = C_1 \phi_1(t - t_0) + \cdots + C_n \phi_n(t - t_0) + \int_{t_0}^t h(t - s)\, dW(s)$

is such that

(35) $\quad X(t_0) = C_1, \ldots, X^{(n-1)}(t_0) = C_n,$

and hence (33) holds. This is the unique process satisfying (24) and taking on the initial conditions specified by (35).

Let $c_1, c_2, \ldots, c_n$ be n real constants. Then the solution to (24) on $[0, \infty)$ having the initial conditions

$$X(0) = c_1, \ldots, X^{(n-1)}(0) = c_n$$

is

$$X(t) = c_1\phi_1(t) + \cdots + c_n\phi_n(t) + \int_0^t h(t-s)\, dW(s).$$

Thus $X(t)$ is normally distributed with mean

(36) $$EX(t) = c_1\phi_1(t) + \cdots + c_n\phi_n(t)$$

and variance

(37) $$\text{Var}(X(t)) = \sigma^2 \int_0^t h^2(t-s)\, ds = \sigma^2 \int_0^t h^2(s)\, ds.$$

Suppose now that the left side of (24) is stable. Then

(38) $$X_0(t) = \int_{-\infty}^t h(t-s)\, dW(s)$$

is well defined (except on a set of probability zero which we can ignore) and satisfies (24) on $(-\infty, \infty)$. This process is a Gaussian process that has zero means and covariance function

$$r_{X_0}(s, t) = \sigma^2 \int_{-\infty}^{\min(s,t)} h(s-u)h(t-u)\, du$$

$$= \sigma^2 \int_{-\infty}^{\infty} h(s-u)h(t-u)\, du.$$

Thus the process is a second order stationary process and

(39) $$r_{X_0}(t) = \sigma^2 \int_{-\infty}^{\infty} h(-u)h(t-u)\, du, \quad -\infty < t < \infty.$$

(We will find $r_{X_0}(t)$ explicitly for $n = 2$ in Section 6.2.1.) The general solution to (24) on an interval can be written in terms of this process as

(40) $$X(t) = C_1\phi_1(t) + \cdots + C_n\phi_n(t) + X_0(t),$$

where $C_1, \ldots, C_n$ are arbitrary random variables. Since $\phi_1(t), \ldots, \phi_n(t)$ all approach zero as $t \to \infty$, it follows from (40) that

$$\lim_{t \to \infty} (X(t) - X_0(t)) = 0 \quad \text{with probability one.}$$

Consider a process $X(t)$, $-\infty < t < \infty$, of the form (40). This is a second order stationary process if and only if $C_1, \ldots, C_n$ each equal zero with probability one. Thus the $X_0(t)$ process given by (38) is the unique second order stationary process that satisfies (24) on $(-\infty, \infty)$.

Let $X(t)$ be a second order process that satisfies (24) on $[0, \infty)$, where the left side of (24) is stable. Then this process can be represented as in (40), where each of the random variables $C_1, \ldots, C_n$ has finite second moment. It follows easily that

(41) $$\lim_{t \to +\infty} E(X(t) - X_0(t))^2 = 0,$$

(42) $$\lim_{t \to +\infty} \mu_X(t) = 0,$$

and

(43) $$\lim_{s,t \to +\infty} (r_X(s, t) - r_{X_0}(s, t)) = 0.$$

In other words, any second order process that satisfies (24) on $[0, \infty)$ is asymptotically equal to a second order stationary process having zero means and covariance function given by (39).

We can also consider stochastic differential equations of the form

(44) $$a_0 X^{(n)}(t) + a_1 X^{(n-1)}(t) + \cdots + a_n X(t) = Y(t),$$

where $Y(t)$, $-\infty < t < \infty$, is a second order stationary process. By a solution to (44) on an interval, we mean an n times differentiable second order process $X(t)$ satisfying (44) on that interval. The results for solutions to (24) extend almost verbatim to solutions to (44) if we replace integrals of the form

$$\int_a^b h(t - s) \, dW(s) \quad \text{by} \quad \int_a^b h(t - s) Y(s) \, ds,$$

except that formulas for the mean and covariance functions of the $X(t)$ process are different.

In particular, if the left side of (44) is stable, the unique second order stationary process that satisfies (44) on $(-\infty, \infty)$ is given by

$$X_0(t) = \int_{-\infty}^t h(t - s) Y(s) \, ds = \int_{-\infty}^\infty h(t - s) Y(s) \, ds.$$

The covariance function of this process is

$$r_{X_0}(s, t) = \int_{-\infty}^\infty \left(\int_{-\infty}^\infty h(s - u) h(t - v) r_Y(v - u) \, dv \right) du$$

or

(45) $$r_{X_0}(t) = \int_{-\infty}^\infty \left(\int_{-\infty}^\infty h(-u) h(t - v) r_Y(v - u) \, dv \right) du.$$

The mean function of this process can be obtained by observing that if (44) holds, then

(46) $\quad a_0\mu_{X_0}^{(n)}(t) + a_1\mu_{X_0}^{(n-1)}(t) + \cdots + a_n\mu_{X_0}(t) = \mu_Y(t).$

Since $X_0(t)$ and $Y(t)$ are second order stationary processes, $\mu_{X_0}(t)$ and $\mu_Y(t)$ take on constant values μ_{X_0} and μ_Y respectively. Thus

$$\mu_{X_0}^{(n)}(t) = \cdots = \mu'_{X_0}(t) = 0,$$

so we conclude from (46) that

(47) $\quad \mu_{X_0} = \dfrac{\mu_Y}{a_n}.$

If $Y(t)$ is a Gaussian process, then so is the $X_0(t)$ process.

Finally we can combine the above results in a fairly obvious manner by considering solutions to the stochastic differential equation

$$a_0 X^{(n)}(t) + a_1 X^{(n-1)}(t) + \cdots + a_n X(t) = Y(t) + W'(t).$$

In particular, in the stable case, the stationary solution to the stochastic differential equation

$$a_0 X^{(n)}(t) + a_1 X^{(n-1)}(t) + \cdots + a_n X(t) = c + W'(t),$$

where c is constant, is given by

$$X_0(t) = \frac{c}{a_n} + \int_{-\infty}^{t} h(t-s)\, dW(s).$$

This process has constant mean c/a_n and covariance function given by (39). We can regard this setup as a model for an input-output system in which the input has signal c and noise $W'(t)$. Then the output has signal c/a_n and noise

$$\int_{-\infty}^{t} h(t-s)\, dW(s).$$

6.2.1. The case $n = 2$. We will now fill in some of the details in the above discussion for $n = 2$. In this case (24) becomes

(48) $\quad a_0 X''(t) + a_1 X'(t) + a_2 X(t) = W'(t),$

whose integrated form on an interval containing the origin is

(49) $\quad a_0(X'(t) - X'(0)) + a_1(X(t) - X(0)) + a_2 \int_0^t X(s)\, ds = W(t).$

The corresponding homogeneous differential equation is

$$a_0 x''(t) + a_1 x'(t) + a_2 x(t) = 0,$$

6.2. Differential equations of order n

and the characteristic polynomial is

$$p(r) = a_0 r^2 + a_1 r + a_2.$$

For completeness we will give formulas for ϕ_1 and ϕ_2 and show how these formulas are derived.

Let us try to find a solution to the homogeneous equation of the form $\phi(t) = e^{rt}$ for some complex constant r. For this choice of $\phi(t)$ we find that

$$a_0 \phi''(t) + a_1 \phi'(t) + a_2 \phi(t) = a_0 r^2 e^{rt} + a_1 r e^{rt} + a_2 e^{rt}$$
$$= (a_0 r^2 + a_1 r + a_2) e^{rt}$$
$$= p(r) e^{rt}.$$

Thus $\phi(t) = e^{rt}$ satisfies the homogeneous equation if and only if $p(r) = 0$, i.e., if and only if r is a root of the characteristic polynomial.

In order to obtain specific formulas for $\phi_1(t)$ and $\phi_2(t)$ we must distinguish three separate cases corresponding to positive, negative, and zero values of the discriminant of the characteristic polynomial.

Case 1. $a_1^2 - 4 a_0 a_2 > 0.$

The characteristic polynomial has two distinct real roots

$$r_1 = \frac{-a_1 + \sqrt{a_1^2 - 4 a_0 a_2}}{2 a_0} \quad \text{and} \quad r_2 = \frac{-a_1 - \sqrt{a_1^2 - 4 a_0 a_2}}{2 a_0}.$$

The functions $e^{r_1 t}$ and $e^{r_2 t}$ are solutions to the homogeneous equation as is any linear combination $c_1 e^{r_1 t} + c_2 e^{r_2 t}$, where c_1 and c_2 are constants. We now choose c_1 and c_2 so that the solution

$$\phi_1(t) = c_1 e^{r_1 t} + c_2 e^{r_2 t}$$

satisfies the initial conditions $\phi_1(0) = 1$ and $\phi_1'(0) = 0$. Since

$$\phi_1'(t) = c_1 r_1 e^{r_1 t} + c_2 r_2 e^{r_2 t},$$

we obtain two equations in the two unknowns c_1 and c_2, namely,

$$c_1 + c_2 = 1$$
$$c_1 r_1 + c_2 r_2 = 0,$$

which have the unique solution

$$c_1 = -\frac{r_2}{r_1 - r_2} \quad \text{and} \quad c_2 = \frac{r_1}{r_1 - r_2}.$$

Thus
$$\phi_1(t) = \frac{r_1 e^{r_2 t} - r_2 e^{r_1 t}}{r_1 - r_2}.$$

By similar methods we find that the solution ϕ_2 to the homogeneous equation having initial conditions $\phi_2(0) = 0$ and $\phi_2'(0) = 1$ is
$$\phi_2(t) = \frac{e^{r_1 t} - e^{r_2 t}}{r_1 - r_2}.$$

Case 2. $\quad a_1^2 - 4a_0 a_2 < 0$.

The characteristic polynomial has two distinct complex-valued roots
$$r_1 = \frac{-a_1 + i\sqrt{4a_0 a_2 - a_1^2}}{2a_0} \quad \text{and} \quad r_2 = \frac{-a_1 - i\sqrt{4a_0 a_2 - a_1^2}}{2a_0}.$$

In terms of r_1 and r_2 the functions $\phi_1(t)$ and $\phi_2(t)$ are given by the same formulas as in Case 1. Alternatively, using the formula $e^{i\theta} = \cos\theta + i\sin\theta$ and elementary algebra, we can rewrite these formulas as
$$\phi_1(t) = e^{\alpha t}\left(\cos\beta t - \frac{\alpha}{\beta}\sin\beta t\right)$$
and
$$\phi_2(t) = \frac{1}{\beta} e^{\alpha t} \sin\beta t,$$
where α and β are real numbers defined by $r_1 = \alpha + i\beta$ or
$$\alpha = -\frac{a_1}{2a_0} \quad \text{and} \quad \beta = \frac{\sqrt{4a_0 a_2 - a_1^2}}{2a_0}.$$

It is clear from these formulas that $\phi_1(t)$ and $\phi_2(t)$ are real-valued functions.

Case 3. $\quad a_1^2 - 4a_0 a_2 = 0$.

The characteristic polynomial has the unique real root
$$r_1 = -\frac{a_1}{2a_0}.$$

One solution to the homogeneous equation is $\phi(t) = e^{r_1 t}$. A second such solution is $\phi(t) = t e^{r_1 t}$:

$$a_0 \phi''(t) + a_1 \phi'(t) + a_2 \phi(t)$$
$$= a_0(r_1^2 t + 2r_1)e^{r_1 t} + a_1(r_1 t + 1)e^{r_1 t} + a_2 t e^{r_1 t}$$
$$= (a_0 r_1^2 + a_1 r_1 + a_2)t e^{r_1 t} + (2a_0 r_1 + a_1)e^{r_1 t}$$
$$= 0.$$

6.2. Differential equations of order n

Thus $\phi_1(t) = c_1 e^{r_1 t} + c_2 t e^{r_1 t}$ is a solution to the homogeneous equation for arbitrary constants c_1 and c_2. Choosing c_1 and c_2 so that $\phi_1(0) = 1$ and $\phi'_1(0) = 0$, we find that

$$\phi_1(t) = e^{r_1 t}(1 - r_1 t).$$

Similarly the solution ϕ_2 satisfying the initial conditions $\phi_2(0) = 0$ and $\phi'_2(0) = 1$ is found to be

$$\phi_2(t) = t e^{r_1 t}.$$

Suppose that the left side of (48) is stable. Then the stationary solution $X_0(t)$ to (48) on $(-\infty, \infty)$ has the covariance function given by (39). Since

$$h(t) = \begin{cases} \dfrac{1}{a_0} \phi_2(t), & t \geq 0, \\ 0, & t < 0, \end{cases}$$

we can use our formulas for $\phi_2(t)$ to compute $r_{X_0}(t)$. The indicated integration is straightforward and leads to the result that

(50) $$r_{X_0}(t) = \frac{\sigma^2}{2 a_1 a_2} \phi_1(|t|), \qquad -\infty < t < \infty,$$

in all three cases for $n = 2$. In particular,

(51) $$\operatorname{Var} X_0(t) = r_{X_0}(0) = \frac{\sigma^2}{2 a_1 a_2}, \qquad -\infty < t < \infty.$$

Example 2. Consider the stochastic differential equation

$$X''(t) + 2X'(t) + 2X(t) = W'(t).$$

(a) Suppose $X(t)$, $0 \leq t < \infty$, is the solution to this equation on $[0, \infty)$ having the initial conditions $X(0) = 0$ and $X'(0) = 1$. Find the distribution of $X(t)$ at the first positive time t such that $EX(t) = 0$.

(b) Consider the stationary solution $X_0(t)$, $-\infty < t < \infty$, to this equation on $(-\infty, \infty)$. Find the first positive time t such that $X_0(0)$ and $X_0(t)$ are uncorrelated.

Since $a_1^2 - 4 a_0 a_2 = 4 - 8 = -4 < 0$, Case 2 is applicable. Now

$$\alpha = -\frac{2}{2} = -1 \quad \text{and} \quad \beta = \frac{\sqrt{8 - 4}}{2} = 1.$$

Thus

$$\phi_1(t) = e^{-t}(\cos t + \sin t), \qquad \phi_2(t) = e^{-t} \sin t,$$

and
$$h(t) = \phi_2(t) = e^{-t} \sin t, \quad t \geq 0.$$

The mean and variance of the solution having the initial conditions indicated in (a) are given according to (36) and (37) by

$$EX(t) = \phi_2(t) = e^{-t} \sin t$$

and

$$\text{Var}(X(t)) = \sigma^2 \int_0^t h^2(s)\, ds = \sigma^2 \int_0^t e^{-2s} \sin^2 s \, ds.$$

Evaluating the last integral, we find that

$$\text{Var}(X(t)) = \frac{\sigma^2}{8} [1 + e^{-2t}(\cos 2t - \sin 2t - 2)].$$

The first positive time t such that $EX(t) = 0$ is $t = \pi$. We see that $X(\pi)$ is normally distributed with mean 0 and variance $\sigma^2(1 - e^{-2\pi})/8$. The covariance function of the stationary solution to the differential equation is given, according to (50), by

$$r_{X_0}(t) = \frac{\sigma^2}{8} e^{-|t|}(\cos |t| + \sin |t|).$$

Thus the first positive time t such that $X_0(0)$ and $X_0(t)$ are uncorrelated is $t = 3\pi/4$.

6.3. Estimation theory

In this section we will study problems of the form of estimating a random variable Y by a random variable $\hat{Y}$, where $\hat{Y}$ is required to be defined in terms of a given stochastic process $X(t)$, $t \in T$. In terms of the probability space Ω, we observe a sample function $X(t, \omega)$, $t \in T$, and use this information to construct an estimate $\hat{Y}(\omega)$ of $Y(\omega)$. Estimation theory is concerned with methods for choosing good estimators.

Example 3. Let $X(t)$, $0 \leq t < \infty$, be a second order process and let $0 < t_0 < t_1$. The problem of estimating $X(t_1)$ from $X(t)$, $0 \leq t \leq t_0$, is called a *prediction* problem. We think of t_0 as the present, $t < t_0$ as the past, and $t > t_0$ as the future. A prediction problem, then, involves estimating future values of a stochastic process from its past and present values. In the absence of any general theory one can only use some intuitively reasonable estimates. We could, for example, estimate $X(t_1)$ by the present value $X(t_0)$. If the $X(t)$ process is differentiable, we could estimate $X(t_1)$ by $X(t_0) + (t_1 - t_0)X'(t_0)$.

6.3. Estimation theory

Example 4. Let $S(t)$, $0 \leq t \leq 1$, be a second order process. Let $N(t)$, $0 \leq t \leq 1$, be a second order process independent of the $S(t)$ process and having zero means. Problems of estimating some random variable defined in terms of the $S(t)$ process based on observation of the process $X(t) = S(t) + N(t)$, $0 \leq t \leq 1$, are called *filtering* problems. Thinking of the $S(t)$ process as a "signal" and of the $N(t)$ process as noise, we wish to filter out most of the noise without appreciably distorting the signal. Suppose we want to estimate the $S(t)$ process at some fixed value of t, say $t = \frac{1}{2}$. If the signal varies slowly in time and the noise oscillates rapidly about zero, it might be reasonable to estimate $S(\frac{1}{2})$ by

$$\frac{1}{2\varepsilon} \int_{\frac{1}{2}-\varepsilon}^{\frac{1}{2}+\varepsilon} X(t) \, dt$$

for some suitable ε between 0 and $\frac{1}{2}$.

We have discussed two examples of estimation problems and described some ad hoc estimators. In order to formulate estimation as a precise mathematical problem, we need some criterion to use in comparing the accuracy of possible estimators. We will use mean square error as our measure of accuracy. We will estimate random variables Y having finite second moment by random variables Z also having finite second moment. The *mean square error* of the estimate is $E(Z - Y)^2$. If Z_1 and Z_2 are two estimators of Y such that $E(Z_1 - Y)^2 < E(Z_2 - Y)^2$, then Z_1 is considered to be the better estimator.

In any particular estimation problem we must estimate some random variable Y in terms of a process $X(t)$, $t \in T$. A random variable Z is an allowable estimator only if it is defined in terms of the $X(t)$ process. We may further restrict the allowable estimators by requiring that they depend on the $X(t)$ process in some suitably simple manner. In any case we obtain some collection $\mathcal{M}$ of random variables which we consider to be the allowable estimators. An *optimal estimator* of Y is a random variable $\hat{Y}$ in $\mathcal{M}$ such that

(52) $$E(\hat{Y} - Y)^2 = \min_{Z \in \mathcal{M}} E(Z - Y)^2.$$

The estimators are required to have finite second moment, so that
 (i) if Z is in $\mathcal{M}$, then $EZ^2 < \infty$.
In almost all cases of interest, $\mathcal{M}$ is such that
 (ii) if Z_1 and Z_2 are in $\mathcal{M}$ and a_1 and a_2 are real constants, then $a_1 Z_1 + a_2 Z_2$ is in $\mathcal{M}$.
If condition (ii) holds, then $\mathcal{M}$ is a vector space. To verify that optimal estimators exist, it is usually necessary for $\mathcal{M}$ to be such that

(iii) if $Z_1, Z_2, \ldots$ are in $\mathcal{M}$ and Z is a random variable such that $\lim_{n \to \infty} E(Z_n - Z)^2 = 0$, then Z is in $\mathcal{M}$.

Condition (iii) states that if Z is the mean square limit of random variables in $\mathcal{M}$, then Z is in $\mathcal{M}$. In other words, this condition states that $\mathcal{M}$ is closed under mean square convergence.

Example 5. Linear estimation. Consider a second order process $X(t)$, $t \in T$. Let $\mathcal{M}_0$ be the collection of all random variables that are of the form of a constant plus a finite linear combination of the random variables $X(t)$, $t \in T$. Thus a random variable is in $\mathcal{M}_0$ if and only if it is of the form

$$a + b_1 X(s_1) + \cdots + b_n X(s_n)$$

for some positive integer n, some numbers $s_1, \ldots, s_n$ each in T, and some real numbers $a, b_1, \ldots, b_n$. The collection $\mathcal{M}_0$ satisfies (i) and (ii), but it does not in general satisfy (iii) because certain random variables involving integration or differentiation, e.g., $X'(t_1)$ for some $t_1 \in T$, may be well defined in terms of the $X(t)$ process but not be in $\mathcal{M}_0$. Such random variables, however, can be mean square limits of random variables in $\mathcal{M}_0$ under appropriate conditions, as we saw in Section 5.3. This leads us to consider the collection $\mathcal{M}$ of all random variables which arise as mean square limits of random variables in $\mathcal{M}_0$. Clearly $\mathcal{M}$ contains $\mathcal{M}_0$. It can be shown that $\mathcal{M}$ satisfies conditions (i), (ii), and (iii). Estimation problems involving this choice of $\mathcal{M}$ are called *linear* estimation problems.

Example 6. Nonlinear estimation. Let $X(t)$, $t \in T$, be a second order process as in the previous example. Let $\mathcal{M}_0$ be the collection of all random variables having finite second moment and of the form

$$f(X(s_1), \ldots, X(s_n)),$$

where n ranges over all positive integers, $s_1, \ldots, s_n$ range over T, and f is an arbitrary real-valued function on R^n (subject to a technical condition involving "measurability"). Again $\mathcal{M}_0$ satisfies conditions (i) and (ii) but not necessarily (iii). The larger collection $\mathcal{M}$ of all random variables arising as mean square limits of random variables in $\mathcal{M}_0$ satisfies all three conditions. Estimation problems involving this choice of $\mathcal{M}$ are called *nonlinear* estimation problems.

The extension from $\mathcal{M}_0$ to $\mathcal{M}$ in the above two examples is necessary only if the parameter set T is infinite. If T is a finite set, then $\mathcal{M}_0 = \mathcal{M}$ in these examples.

6.3.1. General principles of estimation.
Most methods for finding optimal estimators are based on the following theorem.

Theorem 2 *Let $\mathcal{M}$ satisfy conditions* (i) *and* (ii). *Then $\hat{Y} \in \mathcal{M}$ is an optimal estimator of Y if and only if*

(53) $$E(\hat{Y} - Y)Z = 0, \quad Z \in \mathcal{M}.$$

If $\hat{Y}$ and $\hat{\hat{Y}}$ are both optimal estimators of Y, then $E(\hat{\hat{Y}} - \hat{Y})^2 = 0$ and hence $\hat{\hat{Y}} = \hat{Y}$ with probability one; in this sense the optimal estimator of Y is uniquely determined.

Two random variables Z_1 and Z_2, each having finite second moment, are said to be *orthogonal* to each other if $EZ_1Z_2 = 0$. Theorem 2 asserts that an optimal estimator of Y in terms of a random variable lying in $\mathcal{M}$ is the unique random variable $\hat{Y}$ in $\mathcal{M}$ such that $\hat{Y} - Y$ is orthogonal to all the random variables lying in $\mathcal{M}$ (see Figure 2).

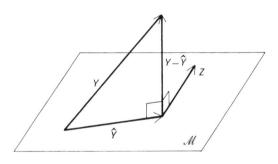

Figure 2

Proof. Let $\hat{Y} \in \mathcal{M}$ be an optimal estimator of Y and let Z be in $\mathcal{M}$. Then by condition (ii), $\hat{Y} + aZ$ is in $\mathcal{M}$. It follows from (52) that

$$E(\hat{Y} - Y)^2 \leq E(\hat{Y} + aZ - Y)^2, \quad -\infty < a < \infty.$$

In other words the function f defined by

$$\begin{aligned} f(a) &= E(\hat{Y} + aZ - Y)^2 \\ &= E(\hat{Y} - Y)^2 + 2aE(\hat{Y} - Y)Z + a^2 EZ^2 \end{aligned}$$

has a minimum at $a = 0$. Thus

$$0 = f'(0) = 2E(\hat{Y} - Y)Z,$$

which shows that (53) holds.

Suppose now that $\hat{Y} \in \mathcal{M}$ and (53) holds. Let $\overset{\ast}{\hat{Y}}$ be any random variable in $\mathcal{M}$. Then

$$E(\overset{\ast}{\hat{Y}} - Y)^2 = E(\hat{Y} - Y + \overset{\ast}{\hat{Y}} - \hat{Y})^2$$
$$= E(\hat{Y} - Y)^2 + 2E(\hat{Y} - Y)(\overset{\ast}{\hat{Y}} - \hat{Y}) + E(\overset{\ast}{\hat{Y}} - \hat{Y})^2.$$

Since $\overset{\ast}{\hat{Y}} - \hat{Y}$ is in $\mathcal{M}$, we can apply (53) with $Z = \overset{\ast}{\hat{Y}} - \hat{Y}$ to conclude that $E(\hat{Y} - Y)(\overset{\ast}{\hat{Y}} - \hat{Y}) = 0$, and hence that

(54) $$E(\overset{\ast}{\hat{Y}} - Y)^2 = E(\hat{Y} - Y)^2 + E(\overset{\ast}{\hat{Y}} - \hat{Y})^2.$$

Since $E(\overset{\ast}{\hat{Y}} - \hat{Y})^2 \geq 0$, (54) shows that $\hat{Y}$ is at least as good an estimator of Y as is $\overset{\ast}{\hat{Y}}$. Since $\overset{\ast}{\hat{Y}}$ is an arbitrary random variable in $\mathcal{M}$, $\hat{Y}$ is an optimal estimator of Y. If $\overset{\ast}{\hat{Y}}$ is also an optimal estimator of Y, then by (54) we see that $E(\overset{\ast}{\hat{Y}} - \hat{Y})^2 = 0$. This completes the proof of the theorem. ∎

It can be shown that if $\mathcal{M}$ satisfies condition (iii) as well as (i) and (ii), then there is always an optimal estimator of Y.

Let $X(t)$, $t \in T$, be a second order process and let $\mathcal{M}$ be as in Example 5. Let $\hat{Y}$ be the optimal linear estimator of a random variable Y. Since the constant random variable $Z = 1$ is in $\mathcal{M}_0$ and hence in $\mathcal{M}$, it follows from (53) that

(55) $$E(\hat{Y} - Y) = 0.$$

Since the random variable $X(t)$ is in $\mathcal{M}_0 \subseteq \mathcal{M}$ for $t \in T$,

(56) $$E(\hat{Y} - Y)X(t) = 0, \qquad t \in T.$$

Conversely, if $\hat{Y} \in \mathcal{M}$ satisfies (55) and (56), then $\hat{Y}$ is the optimal linear estimator of Y. The proof of this result is left as an exercise.

Let $X(t)$, $t \in T$, be a second order process and let Y be a random variable as before. Suppose now that for every positive integer n and every choice of $s_1, \ldots, s_n$ all in T, the random variables $X(s_1), \ldots, X(s_n)$, Y have a joint normal distribution. It can be shown that in this case the optimal linear estimator of Y and the optimal nonlinear estimator of Y coincide. The proof depends basically on the fact that if $X(s_1), \ldots, X(s_n)$, Y have a joint normal distribution, then

$$E[Y \mid X(s_1) = x_1, \ldots, X(s_n) = x_n] = a + b_1 x_1 + \cdots + b_n x_n$$

for suitable constants $a, b_1, \ldots, b_n$.

6.3.2. Some examples of optimal prediction. We will close this section by discussing some examples of prediction problems in which the optimal predictor takes on a particularly simple form.

6.3. Estimation theory

Example 7. Let $W'(t)$ represent white noise with parameter σ^2 and let the observed process $X(t)$, $0 \leq t < \infty$, be the solution to the differential equation

(57) $\quad a_0 X''(t) + a_1 X'(t) + a_2 X(t) = W'(t), \qquad 0 \leq t < \infty,$

satisfying the deterministic initial conditions

$$X(0) = x_0 \quad \text{and} \quad X'(0) = v_0.$$

Let $0 < t_1 < t_2$. Find the optimal linear predictor of $X(t_2)$ in terms of $X(t)$, $0 \leq t \leq t_1$, and find the mean square error of prediction.

As we saw in Section 6.2, we can write the solution to (57) as

(58) $\quad X(t) = X(0)\phi_1(t) + X'(0)\phi_2(t)$

$$+ \int_0^t h(t-s)\,dW(s), \qquad 0 \leq t < \infty,$$

where ϕ_1, ϕ_2, and h are defined explicitly in Section 6.2.1. We have similarly that

$$X(t) = X(t_1)\phi_1(t - t_1) + X'(t_1)\phi_2(t - t_1)$$

$$+ \int_{t_1}^t h(t-s)\,dW(s), \qquad t_1 \leq t < \infty.$$

Set

$$\hat{X}(t_2) = X(t_1)\phi_1(t_2 - t_1) + X'(t_1)\phi_2(t_2 - t_1).$$

Then

$$\hat{X}(t_2) - X(t_2) = -\int_{t_1}^{t_2} h(t_2 - s)\,dW(s).$$

We will show that $\hat{X}(t_2)$ is the optimal linear predictor of $X(t_2)$ in terms of $X(t)$, $0 \leq t \leq t_1$.

We note first that

$$E(\hat{X}(t_2) - X(t_2)) = -E\int_{t_1}^{t_2} h(t_2 - s)\,dW(s) = 0.$$

By (41) of Chapter 5

$$E\left[\int_0^t h(t-s)\,dW(s)\int_{t_1}^{t_2} h(t_2 - s)\,dW(s)\right] = 0, \qquad 0 \leq t \leq t_1.$$

Using (58) and the fact that $X(0)$ and $X'(0)$ have the respective deterministic values x_0 and v_0, we now conclude that for $0 \leq t \leq t_1$

$$E[X(t)(\hat{X}(t_2) - X(t_2))] = E\left[\left(x_0\phi_1(t) + v_0\phi_2(t) + \int_0^t h(t-s)\,dW(s)\right)\right.$$

$$\left.\times \left(-\int_{t_1}^{t_2} h(t_2 - s)\,dW(s)\right)\right] = 0.$$

Thus, to show that $\hat{X}(t_2)$ is the optimal linear predictor of $X(t_2)$ in terms of $X(t)$, $0 \leq t \leq t_1$, it is enough to show that $\hat{X}(t_2)$ is the limit in mean square of linear combinations of the random variables $X(t)$, $0 \leq t \leq t_1$. To do this we need only show that $X'(t_1)$ is such a limit in mean square. But from Equation (24) of Chapter 5 we see that $X'(t_1)$ is the limit in mean square of the random variables

$$\frac{X(t_1) - X\left(t_1 - \frac{1}{n}\right)}{\frac{1}{n}}$$

as $n \to +\infty$. This concludes the proof that $\hat{X}(t_2)$ is the desired optimal predictor of $X(t_2)$.

The mean square error of the predictor $\hat{X}(t_2)$ is

$$E(\hat{X}(t_2) - X(t_2))^2 = E\left(\int_{t_1}^{t_2} h(t_2 - s)\, dW(s)\right)^2$$
$$= \sigma^2 \int_{t_1}^{t_2} h^2(t_2 - s)\, ds$$

or

$$E(\hat{X}(t_2) - X(t_2))^2 = \sigma^2 \int_0^{t_2 - t_1} h^2(s)\, ds.$$

There are several worthwhile observations to be made concerning this example. First, $\hat{X}(t)$, $t \geq t_1$, can be uniquely defined as that function which satisfies the homogeneous equation

$$a_0 \hat{X}''(t) + a_1 \hat{X}'(t) + a_2 \hat{X}(t) = 0, \qquad t \geq t_1,$$

and the initial conditions

$$\hat{X}(t_1) = X(t_1) \quad \text{and} \quad \hat{X}'(t_1) = X'(t_1).$$

Secondly, the mean square error of prediction depends only on the distance between t_1 and t_2 and is an increasing function of that distance. Let ε be any positive number less than t_1. Then the predictor $\hat{X}(t_2)$ is the limit in mean square of linear combinations of the random variables $X(t)$, $t_1 - \varepsilon \leq t \leq t_1$. Thus in predicting $X(t_2)$ in terms of $X(t)$, $0 \leq t \leq t_1$, we need only observe $X(t)$, $t_1 - \varepsilon \leq t \leq t_1$, for an arbitrary small positive number ε. Finally, since the $X(t)$ process is a Gaussian process, the optimal linear predictor $\hat{X}(t_2)$ of $X(t_2)$ in terms of $X(t)$, $0 \leq t \leq t_1$, is also the optimal nonlinear predictor.

The results of Example 7 are readily extended to prediction of stochastic processes defined as solutions to differential equations of order n having white noise inputs. Suppose that $X(t)$, $t \geq 0$, is defined by requiring that

$$(59) \qquad a_0 X^{(n)}(t) + \cdots + a_n X(t) = W'(t)$$

on $0 \leq t < \infty$, and that $X(0), \ldots, X^{(n-1)}(0)$ take on n respective deterministic values. Let $\phi_1, \ldots, \phi_n$ and h be as in Section 6.2. Then for $0 < t_1 < t_2$, the optimal (linear or nonlinear) predictor $\hat{X}(t_2)$ of $X(t_2)$ given $X(t)$, $0 \leq t \leq t_1$, is given by

$$(60) \qquad \hat{X}(t_2) = X(t_1)\phi_1(t_2 - t_1) + \cdots + X^{(n-1)}(t_1)\phi_n(t_2 - t_1).$$

The corresponding function $\hat{X}(t)$, $t \geq t_1$, is the unique function that satisfies the homogeneous equation

$$(61) \qquad a_0 \hat{X}^{(n)}(t) + \cdots + a_n \hat{X}(t) = 0, \qquad t \geq t_1,$$

and the initial conditions

$$(62) \qquad \hat{X}(t_1) = X(t_1), \ldots, \hat{X}^{(n-1)}(t_1) = X^{(n-1)}(t_1).$$

The mean square error of prediction is given by

$$(63) \qquad E(\hat{X}(t_2) - X(t_2))^2 = \sigma^2 \int_0^{t_2 - t_1} h^2(s)\, ds.$$

Suppose now that the left side of (59) is stable and let $X(t)$, $-\infty < t < \infty$, be the stationary solution to (59) on $(-\infty, \infty)$. Then for $-\infty < t_1 < t_2$, the optimal (linear or nonlinear) predictor $\hat{X}(t_2)$ of $X(t_2)$ in terms of $X(t)$, $-\infty < t \leq t_1$, is again given by (60) or (61)–(62), and (63) remains valid.

6.4. Spectral distribution

Let $X(t)$, $-\infty < t < \infty$, be a second order stationary process whose covariance function is such that

$$\int_{-\infty}^{\infty} |r_X(t)|\, dt < \infty.$$

The *spectral density function* $f_X(\lambda)$, $-\infty < \lambda < \infty$, is defined by

$$(64) \qquad f_X(\lambda) = \frac{1}{2\pi} \int_{-\infty}^{\infty} e^{-i\lambda t} r_X(t)\, dt, \qquad -\infty < \lambda < \infty.$$

Techniques involving spectral densities are widely used in estimation problems involving second order stationary processes. Though these techniques are often easy to implement, a proper understanding of them requires material such as complex variable theory that would take us too far afield to discuss. In this section we will simply discuss some elementary properties of spectral densities, find them explicitly in a few special cases, and introduce the more general concept of a spectral distribution function.

Since $r_X(-t) = r_X(t)$, $\sin \lambda t\, r_X(t)$ is an odd function of t, and hence

$$\int_{-\infty}^{\infty} \sin \lambda t\, r_X(t)\, dt = 0.$$

Recalling that $e^{-i\lambda t} = \cos \lambda t - i \sin \lambda t$, we conclude from (64) that

$$f_X(\lambda) = \frac{1}{2\pi} \int_{-\infty}^{\infty} \cos \lambda t\, r_X(t)\, dt, \qquad -\infty < \lambda < \infty.$$

It is clear that f_X is a real-valued function which is symmetric about the origin, i.e., $f_X(-\lambda) = f_X(\lambda)$. Using the fact that covariance functions are nonnegative definite and approximating integrals by sums, one can show that f_X is a nonnegative function. It is also possible to show that f_X is integrable on $(-\infty, \infty)$ and that r_X is given in terms of f_X by the Fourier transform

$$r_X(t) = \int_{-\infty}^{\infty} e^{i\lambda t} f_X(\lambda)\, d\lambda, \qquad -\infty < t < \infty.$$

Since f_X is symmetric about the origin, this reduces to

$$r_X(t) = \int_{-\infty}^{\infty} \cos \lambda t\, f_X(\lambda)\, d\lambda.$$

In particular,

$$\text{Var } X(t) = r_X(0) = \int_{-\infty}^{\infty} f_X(\lambda)\, d\lambda.$$

The function F_X, defined by

$$F_X(\lambda) = \int_{-\infty}^{\lambda} f_X(u)\, du, \qquad -\infty < \lambda < \infty,$$

is called the *spectral distribution function* of the process. It is not a probability distribution function and f_X is not a probability density function unless $r_X(0) = 1$.

Example 8. Let $X(t)$, $-\infty < t < \infty$, be the process from Example 2 of Chapter 4. Its covariance function is of the form

$$r_X(t) = \alpha e^{-\beta |t|},$$

where α and β are suitable positive constants. Thus

$$f_X(\lambda) = \frac{\alpha}{2\pi} \int_{-\infty}^{\infty} e^{-i\lambda t} e^{-\beta |t|}\, dt.$$

6.4. Spectral distribution

Now

$$\int_0^\infty e^{-i\lambda t} e^{-\beta t}\, dt = \int_0^\infty e^{-(\beta + i\lambda)t}\, dt$$

$$= \frac{e^{-(\beta + i\lambda)t}}{-(\beta + i\lambda)}\bigg|_0^\infty = \frac{1}{\beta + i\lambda},$$

since

$$\lim_{t \to \infty} |e^{-(\beta + i\lambda)t}| = \lim_{t \to \infty} e^{-\beta t}|e^{-i\lambda t}|$$

$$= \lim_{t \to \infty} e^{-\beta t} = 0.$$

Similarly,

$$\int_{-\infty}^0 e^{-i\lambda t} e^{\beta t}\, dt = \frac{1}{\beta - i\lambda}.$$

Therefore

$$f_X(\lambda) = \frac{\alpha}{2\pi}\left(\frac{1}{\beta + i\lambda} + \frac{1}{\beta - i\lambda}\right) = \frac{\alpha\beta}{\pi(\beta^2 + \lambda^2)}.$$

Consider a stochastic differential equation

$$a_0 X^{(n)}(t) + a_1 X^{(n-1)}(t) + \cdots + a_n X(t) = Y(t), \qquad -\infty < t < \infty,$$

whose input is a second order stationary process and whose left side is stable. As we saw in Section 6.2, the stationary solution to this differential equation is given by

$$X(t) = \int_{-\infty}^\infty h(t - s) Y(s)\, ds,$$

where h is the impulse response function. The covariance function of the solution is

(65) $$r_X(t) = \int_{-\infty}^\infty \left(\int_{-\infty}^\infty h(-u) h(t - v) r_Y(v - u)\, dv\right) du.$$

The function h is such that

$$\int_{-\infty}^\infty |h(t)|\, dt < \infty.$$

Suppose that

$$\int_{-\infty}^\infty |r_Y(t)|\, dt < \infty.$$

It then follows easily from (65) that

$$\int_{-\infty}^\infty |r_X(t)|\, dt \leq \left(\int_{-\infty}^\infty |h(t)|\, dt\right)^2 \int_{-\infty}^\infty |r_Y(t)|\, dt < \infty.$$

Consequently the $Y(t)$ process and the $X(t)$ process both have spectral density functions. To find the relationship between these two spectral density functions, we first define the *frequency response function* H by

$$H(\lambda) = \int_{-\infty}^{\infty} e^{-i\lambda t} h(t)\, dt, \qquad -\infty < \lambda < \infty.$$

We will show that

(66) $\qquad f_X(\lambda) = f_Y(\lambda)|H(\lambda)|^2, \qquad -\infty < \lambda < \infty.$

It is certainly much easier to compute the spectral density of the $X(t)$ process by (66) than to compute the covariance function by (65). Of course, if one is ultimately interested in the covariance function $r_X(t)$, it is necessary either to use (65) or else to compute the Fourier transform

$$r_X(t) = \int_{-\infty}^{\infty} e^{it\lambda} f_Y(\lambda)|H(\lambda)|^2\, d\lambda.$$

In many cases this Fourier transform can be evaluated easily by using complex variable theory.

Formula (66) follows from (65) in a straightforward manner. We start with

$$f_X(\lambda) = \frac{1}{2\pi} \int_{-\infty}^{\infty} e^{-i\lambda t} r_X(t)\, dt$$

$$= \frac{1}{2\pi} \int_{-\infty}^{\infty}\int_{-\infty}^{\infty}\int_{-\infty}^{\infty} e^{-i\lambda t} h(-u) h(t-v) r_Y(v-u)\, dt\, dv\, du.$$

We first integrate with respect to t:

$$\int_{-\infty}^{\infty} e^{-i\lambda t} h(t-v)\, dt = e^{-i\lambda v}\int_{-\infty}^{\infty} e^{-i\lambda(t-v)} h(t-v)\, dt$$

$$= e^{-i\lambda v}\int_{-\infty}^{\infty} e^{-i\lambda t} h(t)\, dt$$

$$= e^{-i\lambda v} H(\lambda).$$

Thus

$$f_X(\lambda) = \frac{H(\lambda)}{2\pi} \int_{-\infty}^{\infty} h(-u) \left(\int_{-\infty}^{\infty} e^{-i\lambda v} r_Y(v-u)\, dv \right) du.$$

Also,

$$\frac{1}{2\pi}\int_{-\infty}^{\infty} e^{-i\lambda v} r_Y(v-u)\, dv = \frac{e^{-i\lambda u}}{2\pi}\int_{-\infty}^{\infty} e^{-i\lambda(v-u)} r_Y(v-u)\, dv$$

$$= \frac{e^{-i\lambda u}}{2\pi}\int_{-\infty}^{\infty} e^{-i\lambda v} r_Y(v)\, dv$$

$$= e^{-i\lambda u} f_Y(\lambda).$$

6.4. Spectral distribution

Consequently,

$$f_X(\lambda) = f_Y(\lambda)H(\lambda) \int_{-\infty}^{\infty} e^{-i\lambda u} h(-u) \, du$$
$$= f_Y(\lambda)H(\lambda) \int_{-\infty}^{\infty} e^{i\lambda u} h(u) \, du$$
$$= f_Y(\lambda)H(\lambda)H(-\lambda).$$

It is left as an exercise for the reader to show that

(67) $$H(\lambda)H(-\lambda) = |H(\lambda)|^2.$$

From this and the preceding result, we obtain (66) as desired.

Of course, in order for (66) to be useful we must be able to compute the frequency response function H. This turns out to be surprisingly easy:

(68) $$H(\lambda) = \frac{1}{a_0(i\lambda)^n + a_1(i\lambda)^{n-1} + \cdots + a_n}, \quad -\infty < \lambda < \infty.$$

We will now prove (68). The impulse response function h is such that if $y(t)$, $-\infty < t < \infty$, is a bounded continuous function and

$$x(t) = \int_{-\infty}^{\infty} h(t - s)y(s) \, ds,$$

then

(69) $$a_0 x^{(n)}(t) + \cdots + a_n x(t) = y(t), \quad -\infty < t < \infty.$$

This is true even if $y(t)$ is a complex-valued function. Choose $-\infty < \lambda < \infty$ and set

$$y(t) = e^{i\lambda t}, \quad -\infty < t < \infty.$$

Then

$$x(t) = \int_{-\infty}^{\infty} h(t - s)y(s) \, ds$$
$$= \int_{-\infty}^{\infty} h(t - s)e^{i\lambda s} \, ds.$$

By setting $u = t - s$ we conclude that

$$x(t) = \int_{-\infty}^{\infty} h(u)e^{i\lambda(t - u)} \, du = e^{i\lambda t} H(\lambda),$$

and hence that

$$x^{(j)}(t) = (i\lambda)^j H(\lambda) e^{i\lambda t}.$$

Substituting this into (69) we find that

$$(a_0(i\lambda)^n + a_1(i\lambda)^{n-1} + \cdots + a_n)H(\lambda)e^{i\lambda t} = e^{i\lambda t},$$

which implies that (68) holds as desired.

Consider again the stable stochastic differential equation

(70) $$a_0 X^{(n)}(t) + a_1 X^{(n-1)}(t) + \cdots + a_n X(t) = W'(t),$$
$$-\infty < t < \infty,$$

where $W'(t)$ is white noise with parameter σ^2. The covariance function of the stationary solution to (70) is given by

(71) $$r_X(t) = \sigma^2 \int_{-\infty}^{\infty} h(-u) h(t-u) \, du.$$

It follows easily from this that

$$\int_{-\infty}^{\infty} |r_X(t)| \, dt \leq \sigma^2 \left(\int_{-\infty}^{\infty} |h(s)| \, ds \right)^2 < \infty,$$

so that the $X(t)$ process has a well defined spectral density function $f_X(\lambda)$. It is left as an exercise for the reader to use (71) to show that

(72) $$f_X(\lambda) = \frac{\sigma^2}{2\pi} |H(\lambda)|^2, \quad -\infty < \lambda < \infty.$$

Since white noise is not a process in the ordinary sense, its spectral density function cannot be simply defined as a Fourier transform of the covariance function. Suppose the spectral density $f_{W'}(\lambda)$ is defined so that, in analogy with (66),

(73) $$f_X(\lambda) = f_{W'}(\lambda) |H(\lambda)|^2, \quad -\infty < \lambda < \infty.$$

From (72) and (73) we find that

(74) $$f_{W'}(\lambda) = \frac{\sigma^2}{2\pi}, \quad -\infty < \lambda < \infty.$$

We consider (74) as defining the spectral density of white noise. From this definition we see that the spectral density function of white noise is constant over all "frequencies" λ. It is this property of white noise that suggests its name.

Let $X(t)$, $-\infty < t < \infty$, be any second order stationary process. A result known as Bochner's Theorem asserts that if $r_X(0) > 0$, the function $r_X(t)/r_X(0)$, $-\infty < t < \infty$, is necessarily the characteristic function of some probability distribution function $G_X(\lambda)$, $-\infty < \lambda < \infty$. The function F_X defined for $r_X(0) > 0$ by

$$F_X(\lambda) = r_X(0) G_X(\lambda), \quad -\infty < \lambda < \infty,$$

and by $F_X(\lambda) = 0$, $-\infty < \lambda < \infty$, if $r_X(0) = 0$, is called the *spectral distribution function* of the process. Since a probability distribution

6.4. Spectral distribution

function is uniquely determined by its characteristic function, it follows that G_X, and hence also the spectral distribution function, is uniquely determined by the covariance function. If F_X is of the form

$$F_X(\lambda) = \int_{-\infty}^{\lambda} f_X(u)\, du, \qquad -\infty < \lambda < \infty,$$

for some nonnegative function f_X, then f_X is called the *spectral density function* of the process; in this case

$$r_X(t) = \int_{-\infty}^{\infty} e^{it\lambda} f_X(\lambda)\, d\lambda, \qquad -\infty < t < \infty.$$

For the benefit of those readers who are familiar with the Stieltjes integral, it should be pointed out that the covariance function can in general be expressed in terms of the spectral distribution function by means of the Stieltjes integral

$$r_X(t) = \int_{-\infty}^{\infty} e^{it\lambda}\, dF_X(\lambda), \qquad -\infty < \lambda < \infty.$$

If

$$\int_{-\infty}^{\infty} |r_X(t)|\, dt < \infty,$$

the definitions given here are equivalent to those given earlier in this section.

Example 9. Let $X(t)$, $-\infty < t < \infty$, be a second order stationary process, such as in Example 1 of Chapter 4, whose covariance function is given by

$$r_X(t) = \sigma^2 \cos \lambda_1 t,$$

where $\lambda_1 > 0$. Find the spectral distribution function of the process.

Suppose first that $\sigma^2 > 0$. Then $r_X(0) = \sigma^2 > 0$ and

$$\frac{r_X(t)}{r_X(0)} = \cos \lambda_1 t$$

is the characteristic function of a random variable which assigns probability $\frac{1}{2}$ to each of the two points $-\lambda_1$ and λ_1. After multiplying the corresponding probability distribution function by $r_X(0) = \sigma^2$, we find that

$$F_X(\lambda) = \begin{cases} 0, & -\infty < \lambda < -\lambda_1, \\ \sigma^2/2, & -\lambda_1 \leq \lambda < \lambda_1, \\ \sigma^2, & \lambda_1 \leq \lambda < \infty. \end{cases}$$

Clearly this formula is also correct if $\sigma^2 = 0$.

Exercises

1 Let X and Y be random variables each having finite second moment. Show that $E(X - Y)^2 \leq 2EX^2 + 2EY^2$. *Hint:* Verify the identity $E(X - Y)^2 + E(X + Y)^2 = 2EX^2 + 2EY^2$.

2 Let m and f be positive constants and let v_0 and x_0 be real constants. The process $V(t)$, $t \geq 0$, defined as the solution to the stochastic differential equation

$$mV'(t) + fV(t) = W'(t), \qquad V(0) = v_0,$$

is known as *Langevin's velocity process*.
(a) Express this velocity process in terms of white noise.
(b) Find its mean and covariance function.

The process $X(t)$, $t \geq 0$, defined as the solution to the stochastic differential equation

$$mX''(t) + fX'(t) = W'(t), \qquad X(0) = x_0, X'(0) = v_0,$$

is called the *Ornstein-Uhlenbeck process*.
(c) Express the Ornstein-Uhlenbeck process in terms of white noise.
(d) Express the Ornstein-Uhlenbeck process in terms of Langevin's velocity process.
(e) Find the mean and variance of the Ornstein-Uhlenbeck process at time t.

3 Let m and f be positive constants and let $V_0(t)$, $-\infty < t < \infty$, be the stationary solution to the stochastic differential equation

$$mV'(t) + fV(t) = W'(t).$$

(a) Express $V_0(t)$ in terms of white noise.
(b) Find its mean and covariance function.
(c) Show directly that

$$\lim_{s,t \to +\infty} (r_V(s, t) - r_{V_0}(s, t)) = 0,$$

where $V(t)$, $t \geq 0$, is from Exercise 2.

(d) Set

$$X_0(t) = \int_0^t V_0(s)\, ds, \qquad t \geq 0.$$

Show that the $X_0(t)$ process satisfies the stochastic differential equation

$$mX''(t) + fX'(t) = W'(t), \qquad t \geq 0.$$

(e) Express $X_0(t)$ in terms of white noise.
(f) Find the mean and variance of $X_0(t)$.

Exercises

4 Is there a stationary solution to the stochastic differential equation

$$a_0 X'(t) + a_1 X(t) = W'(t)$$

on $-\infty < t < \infty$ if $\alpha = -a_1/a_0$ is positive? If so, how can it be expressed in terms of white noise?

5 Let c be a real constant.
  (a) Define precisely what should be meant by a solution to the stochastic differential equation

$$a_0 X'(t) + a_1 X(t) = c + W'(t).$$

  (b) Show that the general solution to this equation on $0 \le t < \infty$ is

$$X(t) = X(0)e^{\alpha t} + \frac{c}{a_0 \alpha}(e^{\alpha t} - 1) + \frac{1}{a_0}\int_0^t e^{\alpha(t-s)}\, dW(s),$$

  where $\alpha = -a_1/a_0$.
  (c) Suppose $\alpha < 0$. Find the stationary solution to the equation on $-\infty < t < \infty$, and show that it is the unique such solution.

6 In each of the following stochastic differential equations find Var $X(t)$, $t \ge 0$, for the solution on $0 \le t < \infty$ having initial conditions $X(0) = 0$ and $X'(0) = 0$; if the left side of the equation is stable, find the covariance function of the stationary solution on $-\infty < t < \infty$.
  (a) $X''(t) + X'(t) \quad\quad\quad\quad = W'(t);$
  (b) $X''(t) + 3X'(t) + 2X(t) = W'(t);$
  (c) $4X''(t) + 8X'(t) + 5X(t) = W'(t);$
  (d) $X''(t) + 2X'(t) + \quad X(t) = W'(t);$
  (e) $X''(t) + \quad\quad\quad\quad\quad X(t) = W'(t).$

7 Show that the left side of the stochastic differential equation

$$a_0 X''(t) + a_1 X'(t) + a_2 X(t) = W'(t)$$

is stable if and only if the coefficients a_0, a_1, and a_2 are either all positive or all negative.

8 Suppose that the left side of the stochastic differential equation

$$a_0 X''(t) + a_1 X'(t) + a_2 X(t) = W'(t)$$

is stable and let $X_0(t)$, $-\infty < t < \infty$, be its stationary solution.
  (a) Show that in Cases 1 and 3 the correlation between $X_0(s)$ and $X_0(t)$ is positive for all s and t.
  (b) Show that in Case 2 there exist choices of s and t such that $X_0(s)$ and $X_0(t)$ are negatively correlated.

9 Let $X_0(t)$, $-\infty < t < \infty$, be the stationary solution to

$$a_0 X''(t) + a_1 X'(t) + a_2 X(t) = W'(t),$$

where the left side of this stochastic differential equation is stable.

(a) Show that $X_0'(t)$ has the covariance function

$$r_{X_0'}(t) = \frac{-\sigma^2}{2a_1 a_2} \phi_1''(|t|).$$

(b) Find $\phi_1''(0)$ and use this to compute Var $X_0'(t)$. *Hint:* Use the definition of $\phi_1(t)$ rather than its explicit formula.

10 Let $X(t)$, $-\infty < t < \infty$, satisfy the stochastic differential equation

$$a_0 X'(t) + a_1 X(t) = W'(t).$$

Let $Y(t)$, $-\infty < t < \infty$, satisfy the stochastic differential equation

$$b_0 Y'(t) + b_1 Y(t) = X(t).$$

Show that the $Y(t)$ process satisfies the stochastic differential equation

$$a_0 b_0 Y''(t) + (a_0 b_1 + a_1 b_0) Y'(t) + a_1 b_1 Y(t) = W'(t).$$

11 Let $Y(t)$, $-\infty < t < \infty$, be a second order stationary process having continuous sample functions, mean $\mu_Y = 1$, and covariance function $r_Y(t) = e^{-|t|}$, $-\infty < t < \infty$.
(a) Find the mean and covariance function of the stationary solution $X_0(t)$, $-\infty < t < \infty$, to the stochastic differential equation

$$X'(t) + X(t) = Y(t).$$

(b) Find the mean and covariance functions of the solution $X(t)$, $0 \le t < \infty$, to this stochastic differential equation satisfying the initial condition $X(0) = 0$.
(c) Show directly that

$$\lim_{s,t \to +\infty} (r_X(s, t) - r_{X_0}(s, t)) = 0.$$

12 Let $\mathcal{M}$ be as in Example 5 and suppose that $\hat{Y} \in \mathcal{M}$ satisfies (55) and (56). Show that $\hat{Y}$ is the optimal linear estimator of Y.

13 Let $X(t)$, $-\infty < t < \infty$, be a second order stationary process having mean zero and covariance function $r(t)$, $-\infty < t < \infty$.
(a) Find the optimal predictor of $X(1)$ of the form $\hat{X}(1) = bX(0)$, and determine the mean square error of prediction.
(b) Find the optimal predictor of $X(1)$ of the form $\hat{X}(1) = b_1 X(0) + b_2 X'(0)$, and determine the mean square error of prediction. Assume here that the $X(t)$ process is differentiable.
(c) Find the optimal estimator of $\int_0^1 X(t)\,dt$ of the form $b_1 X(0) + b_2 X(1)$, and determine the mean square error of estimation. Assume here that $|r(1)| < r(0)$.

14 Show that for $t_1 < t_2$ the optimal (linear or nonlinear) predictor of $W(t_2)$ in terms of $W(t)$, $t \le t_1$, is $\hat{W}(t_2) = W(t_1)$.

15 Let $X(t)$, $-\infty < t < \infty$, be a second order stationary process having mean zero and continuous covariance function. Show that the optimal linear predictor of $X(t + s)$ in terms of $X(0)$ and $X(s)$ is the same as the optimal linear predictor of $X(t + s)$ in terms of $X(s)$ for all $s \geq 0$ and $t \geq 0$ if and only if

$$r_X(t) = r_X(0)e^{-\alpha|t|}, \qquad -\infty < t < \infty,$$

for some nonnegative constant α. *Hint:* Use the fact that a bounded continuous real-valued function $f(t), 0 \leq t < \infty$, satisfies the equation

$$f(s + t) = f(s)f(t), \qquad s \geq 0, t \geq 0$$

if and only if

$$f(t) = e^{-\alpha t}, \qquad 0 \leq t < \infty,$$

for some nonnegative constant α.

16 Let $X(t), 0 \leq t < \infty$, be the solution to the stochastic differential equation

$$a_0 X'(t) + a_1 X(t) = W'(t)$$

satisfying the initial condition $X(0) = 0$. Find the optimal (linear or nonlinear) predictor of $X(t_1 + \tau)$ by $X(t), 0 \leq t \leq t_1$, where t_1 and τ are positive constants. Determine the mean square error of prediction.

17 For each of the stochastic differential equations in Exercise 6 let $X(t)$, $0 \leq t < \infty$, be the solution satisfying the initial conditions $X(0) = 0$ and $X'(0) = 0$ (or any other deterministic initial conditions). Find explicitly the optimal (linear or nonlinear) predictor of $X(t_1 + \tau)$ in terms of $X(t), 0 \leq t \leq t_1$, where t_1 and τ are positive constants.

18 Verify Formula (67).

19 Let $Y(t), -\infty < t < \infty$, be a second order stationary process with spectral density $f_Y(\lambda), -\infty < \lambda < \infty$. Set

$$X(t) = \int_{t-\frac{1}{2}}^{t+\frac{1}{2}} Y(s) \, ds, \qquad -\infty < t < \infty.$$

(a) Find a function $h(t), -\infty < t < \infty$, such that

$$X(t) = \int_{-\infty}^{\infty} h(t - s) Y(s) \, ds, \qquad -\infty < t < \infty.$$

(b) Show that $X(t), -\infty < t < \infty$, is a second order stationary process and find its spectral density function.

20 Let $X(t), -\infty < t < \infty$, be a second order stationary process having spectral density f_X and set $Y(t) = X(t + 1) - X(t), -\infty < t < \infty$. Use the formula for the covariance function of the $Y(t)$ process given in Exercise 3 of Chapter 4 to show that Y has the spectral density

$$f_Y(\lambda) = 2(1 - \cos \lambda) f_X(\lambda), \qquad -\infty < \lambda < \infty.$$

21 Let $h(t)$, $-\infty < t < \infty$, be a continuously differentiable function such that

$$\int_{-\infty}^{\infty} |h(t)|\, dt < \infty \quad \text{and} \quad \int_{-\infty}^{\infty} h^2(t)\, dt < \infty.$$

Then the process

$$X(t) = \int_{-\infty}^{\infty} h(t-s)\, dW(s), \quad -\infty < t < \infty,$$

is a second order stationary Gaussian process whose covariance function is given by (71). Show that it has a spectral density function given by (72).

22 Use the result of the previous exercise to compute the spectral density of the process $X(t) = W(t+1) - W(t)$, $-\infty < t < \infty$.

23 Find the spectral density of the process $X_0(t)$, $-\infty < t < \infty$, defined in Exercise 11.

24 Let $X(t)$, $-\infty < t < \infty$, be the stationary solution to

$$a_0 X''(t) + a_1 X'(t) + a_2 X(t) = W'(t),$$

where the left side of this stochastic differential equation is stable. Show that the $X(t)$ process has the spectral density given by

$$f_X(\lambda) = \frac{\sigma^2}{2\pi[(a_0\lambda^2 - a_2)^2 + a_1^2\lambda^2]}.$$

25 A transmitter transmits a constant but unknown signal s. The output of a receiver is the stationary solution $X(t)$, $-\infty < t < \infty$, to the stable stochastic differential equation

$$a_0 X^{(n)}(t) + \cdots + a_n X(t) = s + W'(t).$$

Suppose that s is estimated from $X(t)$, $0 \le t \le T$, by

$$\hat{s} = \frac{a_n}{T} \int_0^T X(t)\, dt.$$

Show that $\hat{s}$ is an unbiased estimator of s (i.e., $E\hat{s} = s$), and that

$$\lim_{T \to \infty} T\, \text{Var}\, \hat{s} = \sigma^2.$$

Hint: Use Exercise 7 of Chapter 5 and observe that

$$\int_{-\infty}^{\infty} r_X(t)\, dt = 2\pi f_X(0).$$

Exercises

26 Let $X(t)$, $-\infty < t < \infty$, be a second order stationary process having a covariance function of the form

$$r_X(t) = \sigma_1^2 \cos \lambda_1 t + \sigma_2^2 \cos \lambda_2 t, \quad -\infty < t < \infty,$$

where $0 < \lambda_1 < \lambda_2$ (see Example 5 of Chapter 4). Find the spectral distribution function of the $X(t)$ process.

Answers

CHAPTER 1

1 (a) $(1-p)^2/[(1-p)^2 + pq]$,
(b) $\pi_0(0)(1-p-q)(p-q) + q(p+1-q)$.

2 $P(x, y) = \begin{cases} (x/d)^2, & y = x - 1, \\ 2x(d-x)/d^2, & y = x, \\ (d-x)^2/d^2, & y = x + 1, \\ 0, & \text{elsewhere.} \end{cases}$

3 $P(x, y) = \begin{cases} f(y), & x = 0, \\ pf(y - x + 1) + (1-p)f(y - x), & x \geq 1. \end{cases}$

5 (a) $P_0(T_0 = 1) = 1 - p$ and $P_0(T_0 = n) = pq(1-q)^{n-2}$, $n \geq 2$.
(b) $p(1-p)^{n-1}$.

6 $P(X_1 = x) = P(X_0 = x)$.

10 (a) $P_x(T_0 = 1) = \begin{cases} \frac{1}{3}, & x = 1, \\ 0, & \text{elsewhere;} \end{cases}$ $P_x(T_0 = 2) = \begin{cases} \frac{1}{3}, & x = 0, \\ \frac{2}{9}, & x = 2, \\ 0, & \text{elsewhere;} \end{cases}$

$P_x(T_0 = 3) = \begin{cases} \frac{4}{27}, & x = 1, \\ \frac{2}{9}, & x = 3, \\ 0, & \text{elsewhere.} \end{cases}$

(b) $P = \begin{bmatrix} 0 & 1 & 0 & 0 \\ \frac{1}{3} & 0 & \frac{2}{3} & 0 \\ 0 & \frac{2}{3} & 0 & \frac{1}{3} \\ 0 & 0 & 1 & 0 \end{bmatrix}$, $P^2 = \begin{bmatrix} \frac{1}{3} & 0 & \frac{2}{3} & 0 \\ 0 & \frac{7}{9} & 0 & \frac{2}{9} \\ \frac{2}{9} & 0 & \frac{7}{9} & 0 \\ 0 & \frac{2}{3} & 0 & \frac{1}{3} \end{bmatrix}$,

and $P^3 = \begin{bmatrix} 0 & \frac{7}{9} & 0 & \frac{2}{9} \\ \frac{7}{27} & 0 & \frac{20}{27} & 0 \\ 0 & \frac{20}{27} & 0 & \frac{7}{27} \\ \frac{2}{9} & 0 & \frac{7}{9} & 0 \end{bmatrix}$.

(c) $\pi_1 = (\frac{1}{12}, \frac{5}{12}, \frac{5}{12}, \frac{1}{12})$, $\pi_2 = (\frac{5}{36}, \frac{13}{36}, \frac{13}{36}, \frac{5}{36})$,
and $\pi_3 = (\frac{13}{108}, \frac{41}{108}, \frac{41}{108}, \frac{13}{108})$.

Answers

11 (a) $P = \begin{array}{c} 0 \\ 1 \\ 2 \\ 3 \end{array} \begin{bmatrix} 1 & 0 & 0 & 0 \\ \frac{1}{5} & \frac{3}{5} & \frac{1}{5} & 0 \\ 0 & \frac{1}{5} & \frac{3}{5} & \frac{1}{5} \\ 0 & 0 & 0 & 1 \end{bmatrix}$ and $P^2 = \begin{array}{c} 0 \\ 1 \\ 2 \\ 3 \end{array} \begin{bmatrix} 1 & 0 & 0 & 0 \\ \frac{8}{25} & \frac{2}{5} & \frac{6}{25} & \frac{1}{25} \\ \frac{1}{25} & \frac{6}{25} & \frac{2}{5} & \frac{8}{25} \\ 0 & 0 & 0 & 1 \end{bmatrix}$.

(b) $\pi_1 = (\frac{1}{10}, \frac{2}{5}, \frac{2}{5}, \frac{1}{10})$ and $\pi_2 = (\frac{9}{50}, \frac{8}{25}, \frac{8}{25}, \frac{9}{50})$.

(c) $P_0(T_{\{0,3\}} = 1) = P_3(T_{\{0,3\}} = 1) = 1$ and
$P_1(T_{\{0,3\}} = 1) = P_2(T_{\{0,3\}} = 1) = \frac{1}{5}$.
$P_0(T_{\{0,3\}} = 2) = P_3(T_{\{0,3\}} = 2) = 0$ and
$P_1(T_{\{0,3\}} = 2) = P_2(T_{\{0,3\}} = 2) = \frac{4}{25}$.

12 (a) $P^2 = \begin{bmatrix} 1-p & 0 & p \\ 0 & 1 & 0 \\ 1-p & 0 & p \end{bmatrix}$, (c) $P^n = \begin{cases} P, & n \text{ odd,} \\ P^2, & n \text{ even.} \end{cases}$

14 $E_x(X_n) = \dfrac{d}{2} + \left(1 - \dfrac{2}{d}\right)^n \left(x - \dfrac{d}{2}\right)$.

18 (b) $p^{n-1}(1-p)$.

19 (a) 0 is transient and all other states are recurrent.
(b) $\rho_{00} = \frac{1}{2}$, $\rho_{01} = \rho_{02} = \rho_{03} = \frac{3}{4}$, and $\rho_{04} = \rho_{05} = \rho_{06} = \frac{1}{4}$.

20 (a) 3 and 5 are transient and the other states are recurrent.
(b) $\rho_{\{0,1\}}(0) = \rho_{\{0,1\}}(1) = 1$, $\rho_{\{0,1\}}(2) = \rho_{\{0,1\}}(4) = 0$, $\rho_{\{0,1\}}(3) = \frac{7}{11}$, and $\rho_{\{0,1\}}(5) = \frac{6}{11}$.

23 $\rho_{\{0\}}(x) = 1 - (x/2d)$, $0 < x < 2d$.

24 $P_x(T_0 < T_d) = [(q/p)^x - (q/p)^d]/[1 - (q/p)^d]$, $0 < x < d$.

25 (a) .1, (b) $99.10.

29 (b) $1 - \dfrac{6}{\pi^2} \sum_{y=0}^{x-1} \dfrac{1}{(y+1)^2}$.

30 (a) $P_x(T_a < T_b) = (a+1)(b-x)/(x+1)(b-a)$, $a < x < b$;
(b) $\rho_{x0} = 1/(x+1)$, $x > 0$.

32 $4(\sqrt{5} - 2)$.

33 $(\sqrt{5} - 1)/2$.

38 (a) 0 is transient and all other states are absorbing (and hence recurrent).
(b) 0 and 1 are recurrent and all other states are transient.
(c) 0 is absorbing and all other states are transient.
(d) All states are transient.

CHAPTER 2

1 $\pi(0) = .3$, $\pi(1) = .4$, and $\pi(2) = .3$.

6 When $p < \frac{1}{2}$, the stationary distribution exists and is given by
$$\pi(0) = \frac{1 - 2p}{2(1-p)} \quad \text{and} \quad \pi(x) = \left(\frac{1-2p}{2}\right) \frac{p^{x-1}}{(1-p)^{x+1}}, \quad x \geq 1.$$

7 (a) $\pi(x) = \binom{d}{x}\frac{1}{2^d}$, $x = 0, 1, \ldots, d$, which is the binomial distribution with parameters d and $\frac{1}{2}$.
(b) mean $d/2$ and variance $d/4$.

8 $P(x, y) = \begin{cases} x/2d, & y = x - 1, \\ \frac{1}{2}, & y = x, \\ (d-x)/2d, & y = x + 1, \\ 0, & \text{elsewhere.} \end{cases}$

9 $\pi(x) = \binom{d}{x}^2 \Big/ \binom{2d}{d}$, $x = 0, 1, \ldots, d$.

13 $(\lambda/q)p^n$.

14 $\pi(x) = (1 - p)p^x$, $x \geq 0$.

15 $\pi(x) = 1/d$, $x \in \mathscr{S}$.

17 2^d.

18 (b) $\pi(x) = \dfrac{1}{2c}$, $1 \leq x \leq c$, and $\pi(x) = \dfrac{1}{2d}$, $c + 1 \leq x \leq c + d$.

19 (a) $\pi_0 = (0, \frac{1}{3}, \frac{1}{3}, \frac{1}{3}, 0, 0, 0)$ and $\pi_1 = (0, 0, 0, 0, \frac{1}{3}, \frac{1}{3}, \frac{1}{3})$.

(b) $\left(\lim_{n \to \infty} \dfrac{G_n(x, y)}{n}\right) = \begin{array}{c} \\ 0 \\ 1 \\ 2 \\ 3 \\ 4 \\ 5 \\ 6 \end{array} \begin{bmatrix} 0 & 1 & 2 & 3 & 4 & 5 & 6 \\ 0 & 0 & \frac{3}{12} & \frac{3}{12} & \frac{3}{12} & \frac{1}{12} & \frac{1}{12} & \frac{1}{12} \\ 0 & \frac{1}{3} & \frac{1}{3} & \frac{1}{3} & 0 & 0 & 0 \\ 0 & \frac{1}{3} & \frac{1}{3} & \frac{1}{3} & 0 & 0 & 0 \\ 0 & \frac{1}{3} & \frac{1}{3} & \frac{1}{3} & 0 & 0 & 0 \\ 0 & 0 & 0 & 0 & \frac{1}{3} & \frac{1}{3} & \frac{1}{3} \\ 0 & 0 & 0 & 0 & \frac{1}{3} & \frac{1}{3} & \frac{1}{3} \\ 0 & 0 & 0 & 0 & \frac{1}{3} & \frac{1}{3} & \frac{1}{3} \end{bmatrix}.$

20 (a) $\pi_0 = (\frac{2}{5}, \frac{3}{5}, 0, 0, 0, 0)$ and $\pi_1 = (0, 0, \frac{6}{13}, 0, \frac{7}{13}, 0)$.

(b) $\left(\lim_{n \to \infty} \dfrac{G_n(x, y)}{n}\right) = \begin{array}{c} \\ 0 \\ 1 \\ 2 \\ 3 \\ 4 \\ 5 \end{array} \begin{bmatrix} 0 & 1 & 2 & 3 & 4 & 5 \\ \frac{2}{5} & \frac{3}{5} & 0 & 0 & 0 & 0 \\ \frac{2}{5} & \frac{3}{5} & 0 & 0 & 0 & 0 \\ 0 & 0 & \frac{6}{13} & 0 & \frac{7}{13} & 0 \\ \frac{14}{55} & \frac{21}{55} & \frac{24}{143} & 0 & \frac{28}{143} & 0 \\ 0 & 0 & \frac{6}{13} & 0 & \frac{7}{13} & 0 \\ \frac{12}{55} & \frac{18}{55} & \frac{30}{143} & 0 & \frac{35}{143} & 0 \end{bmatrix}.$

21 (a) $P_0(X_n = 0) \doteq \frac{1}{8}$, $P_0(X_n = 2) \doteq \frac{3}{4}$, $P_0(X_n = 4) \doteq \frac{1}{8}$, and $P_0(X_n = 1) = P_0(X_n = 3) = 0$.
(b) $P_0(X_n = 1) \doteq \frac{1}{2}$, $P_0(X_n = 3) \doteq \frac{1}{2}$, and $P_0(X_n = 0) = P_0(X_n = 2) = P_0(X_n = 4) = 0$.

22 (b) 1. (c) $\pi = (\frac{2}{5}, \frac{1}{5}, \frac{2}{5})$.

23 (b) 3. (c) $\pi = (\frac{1}{3}, \frac{1}{9}, \frac{2}{9}, \frac{1}{12}, \frac{1}{4})$.

CHAPTER 3

2 $P_{00}(t) = \dfrac{\mu_1}{\lambda_0 + \lambda_1 + \mu_1} + \dfrac{\lambda_1}{\lambda_1 + \mu_1} e^{-\lambda_0 t} + \dfrac{\lambda_0 \mu_1}{(\lambda_0 + \lambda_1 + \mu_1)(\lambda_1 + \mu_1)} e^{-(\lambda_0 + \lambda_1 + \mu_1)t}$

$P_{01}(t) = \dfrac{\lambda_0}{\lambda_0 + \lambda_1 + \mu_1} - \dfrac{\lambda_0}{\lambda_0 + \lambda_1 + \mu_1} e^{-(\lambda_0 + \lambda_1 + \mu_1)t}$

$P_{02}(t) = \dfrac{\lambda_1}{\lambda_0 + \lambda_1 + \mu_1} - \dfrac{\lambda_1}{\lambda_1 + \mu_1} e^{-\lambda_0 t} + \dfrac{\lambda_0 \lambda_1}{(\lambda_0 + \lambda_1 + \mu_1)(\lambda_1 + \mu_1)} e^{-(\lambda_0 + \lambda_1 + \mu_1)t}$

3 $\dbinom{n}{m} \left(\dfrac{s}{t}\right)^m \left(1 - \dfrac{s}{t}\right)^{n-m}$.

4 $F_{T_m}(t) = 1 - e^{-\lambda t} \displaystyle\sum_{n=0}^{m-1} \dfrac{(\lambda t)^n}{n!}$, $t > 0$,   and $F_{T_m}(t) = 0$, $t \le 0$.

5 T_m has the gamma density with parameters m and λ; i.e.,

$$f_{T_m}(t) = \lambda^m t^{m-1} e^{-\lambda t}/\Gamma(m), \quad t > 0, \quad \text{and} \quad f_{T_m}(t) = 0, \quad t \le 0.$$

6 $1 - \left(1 - \dfrac{s}{t}\right)^n$.

7 $P(X(T) = n) = \lambda^n v/(\lambda + v)^{n+1}$.

8 $P(X(T) = n) = \dfrac{\lambda^n v^\alpha \Gamma(n + \alpha)}{(\lambda + v)^{n+\alpha} n! \Gamma(\alpha)}$.

10 (a) $P'_{xy}(t) = -\mu_y P_{xy}(t) + \mu_{y+1} P_{x,y+1}(t)$, $y \le x - 1$;
$P'_{xx}(t) = -\mu_x P_{xx}(t)$.
(b) $P_{xx}(t) = e^{-\mu_x t}$.

(c) $P_{xy}(t) = \mu_{y+1} \displaystyle\int_0^t e^{-\mu_y(t-s)} P_{x,y+1}(s)\, ds$, $y < x$.

(d) $P_{x,x-1}(t) = \dfrac{\mu_x}{\mu_{x-1} - \mu_x} (e^{-\mu_x t} - e^{-\mu_{x-1} t})$, $\mu_{x-1} \ne \mu_x$,

and $P_{x,x-1}(t) = \mu_x t e^{-\mu_x t}$, $\mu_{x-1} = \mu_x$.

11 $EX(t) = \dfrac{\lambda}{\mu}(1 - e^{-\mu t}) + xe^{-\mu t}$, $\text{Var } X(t) = \left(\dfrac{\lambda}{\mu} + xe^{-\mu t}\right)(1 - e^{-\mu t})$.

12 (a) $P'_{xy}(t) = (y - 1)\lambda P_{x,y-1}(t) - y(\lambda + \mu) P_{xy}(t) + (y + 1)\mu P_{x,y+1}(t)$.

13 (b) $s_x(t) = xe^{2(\lambda - \mu)t} \left[x + \dfrac{\lambda + \mu}{\lambda - \mu}(1 - e^{-(\lambda - \mu)t}) \right]$, $\lambda \ne \mu$,

and $s_x(t) = x(x + 2\lambda t)$, $\lambda = \mu$.

(c) $\text{Var } X(t) = x \dfrac{\lambda + \mu}{\lambda - \mu}(e^{2(\lambda - \mu)t} - e^{(\lambda - \mu)t})$, $\lambda \ne \mu$,

and $\text{Var } X(t) = 2x\lambda t$, $\lambda = \mu$.

14 (a) $\mu_x = x\mu$ and $\lambda_x = (d-x)\lambda$.

(b) $P_{xd}(t) = \dfrac{\lambda^{d-x}}{(\lambda + \mu)^d} (\lambda + \mu e^{-(\lambda+\mu)t})^x (1 - e^{-(\lambda+\mu)t})^{d-x}$.

(c) $xe^{-(\lambda+\mu)t} + \dfrac{d\lambda}{\lambda + \mu}(1 - e^{-(\lambda+\mu)t})$.

16 (a) null recurrent. (b) transient.

20 $\pi(x) = \dbinom{d}{x} \left(\dfrac{\lambda}{\lambda+\mu}\right)^x \left(\dfrac{\mu}{\lambda+\mu}\right)^{d-x}$.

21 $\pi(x) = \dfrac{(\lambda/\mu)^x}{x!} \bigg/ \displaystyle\sum_{y=0}^{d} \dfrac{(\lambda/\mu)^y}{y!}$.

22 (b) The average rate at which customers are served equals the arrival rate λ.

CHAPTER 4

5 $\mu_Y(t) = 0$ and $r_Y(s,t) = \lambda(\min(s,t) - st)$.

6 $\mu_X(t) = t$ and $r_X(s,t) = \dfrac{1}{n}(\min(s,t) - st)$.

7 $r_{XY}(s,t) = r_X(t-s+1)$.

12 $\mu_Y(t) = f(t)\mu_X(g(t))$ and $r_Y(s,t) = f(s)f(t)r_X(g(s),g(t))$.

13 (a) $\mu_Y(t) = r_X(0)$ and $r_Y(s,t) = 2(r_X(s,t))^2$.

14 (a) $\dfrac{1}{\sqrt{2\pi\sigma_2^2(1-\rho^2)}} \exp\left[-\dfrac{1}{2(1-\rho^2)}\left(\dfrac{x_2-\mu_2}{\sigma_2} - \rho\dfrac{x_1-\mu_1}{\sigma_1}\right)^2\right]$.

(b) $\mu_2 + \rho\dfrac{\sigma_2}{\sigma_1}(x_1 - \mu_1)$.

17 Normal with mean 0 and variance $\sigma^2 n(n+1)(2n+1)/6$.

20 (a) $\mu_X(t) = \sigma^2 t$ and $r_X(s,t) = 2\sigma^4 \min(s^2, t^2)$.
(b) and (c) $\mu_X(t) = 0$ and $r_X(s,t) = \sigma^2 \min(s,t)$.
(d) $\mu_X(t) = 0$ and $r_X(s,t) = \sigma^2(\min(s,t) - st)$.

CHAPTER 5

2 $\sqrt{3t}\,(2-t)/2$.

3 mean $\sigma^2/2$ and variance $\sigma^4/3$.

4 $\mu_X(t) = 0$; $r_X(s,t) = \dfrac{\sigma^2}{2}s^2\left(t - \dfrac{s}{3}\right)$, $s \le t$; and

$r_X(s,t) = \dfrac{\sigma^2}{2}t^2\left(s - \dfrac{t}{3}\right)$, $s > t$.

Answers

5 $r_X(t) = \dfrac{\sigma^2}{6}(1 - |t|)^2(2 + |t|)$, $\quad |t| \le 1$, $\quad$ and $\quad r_X(t) = 0$, $\quad |t| > 1$.

12 (a) $r_X^{(2)}(t - s)$, $\quad$ (b) $-r_X^{(3)}(t - s)$, $\quad$ (c) $r_X^{(4)}(t - s)$.

13 $\mu_Y(t) = \mu_X$ $\quad$ and $\quad r_Y(t) = r_X^{(4)}(t) + 2r_X^{(2)}(t) + r_X(t)$.

14 $c(W(b) - W(a))$.

15 $EX = 0$, $\quad \text{Var } X = \sigma^2/3$, $\quad EY = 0$, $\quad \text{Var } Y = \sigma^2/5$, $\quad$ and $\quad \rho = \sqrt{15}/4$.

16 (a) $\dfrac{\sigma^2}{3} \min(s^3, t^3)$, $\qquad$ (b) $\dfrac{\sigma^2}{2}\left(\dfrac{\sin(s + t)}{s + t} + \dfrac{\sin(s - t)}{s - t}\right)$,

(c) $\dfrac{\sigma^2}{6}(|t - s| - 1)^2(|t - s| + 2)$, $\quad |s - t| \le 1$, $\quad$ and $\quad$ 0 elsewhere.

17 (b) $\dfrac{\sigma^2}{2\alpha^3}(e^{2\alpha t} - 4e^{\alpha t} + 2\alpha t + 3)$.

CHAPTER 6

2 (a) $V(t) = v_0 e^{\alpha t} + \dfrac{1}{m}\displaystyle\int_0^t e^{\alpha(t-s)}\, dW(s)$, $\quad$ where $\alpha = -f/m$.

(b) $\mu_V(t) = v_0 e^{\alpha t}$, $\quad r_V(s, t) = \dfrac{\sigma^2}{2mf}(e^{\alpha|t-s|} - e^{\alpha(s+t)})$.

(c) $X(t) = x_0 + \dfrac{v_0}{\alpha}(e^{\alpha t} - 1) + \dfrac{1}{m\alpha}\displaystyle\int_0^t (e^{\alpha(t-u)} - 1)\, dW(u)$.

(d) $X(t) = x_0 + \displaystyle\int_0^t V(u)\, du$.

(e) $\mu_X(t) = x_0 + \dfrac{v_0}{\alpha}(e^{\alpha t} - 1)$, $\quad \text{Var } X(t) = \dfrac{\sigma^2 m}{2f^3}(4e^{\alpha t} - e^{2\alpha t} - 2\alpha t - 3)$.

3 (a) $V_0(t) = \dfrac{1}{m}\displaystyle\int_{-\infty}^t e^{\alpha(t-s)}\, dW(s)$.

(b) $\mu_{V_0}(t) = 0$, $\quad r_{V_0}(t) = \dfrac{\sigma^2}{2mf}e^{\alpha|t|}$.

(e) $X_0(t) = \dfrac{e^{\alpha t} - 1}{m\alpha}\displaystyle\int_{-\infty}^0 e^{-\alpha u}\, dW(u) + \dfrac{1}{m\alpha}\displaystyle\int_0^t (e^{\alpha(t-u)} - 1)\, dW(u)$.

(f) $\mu_{X_0}(t) = 0$, $\quad \text{Var } X_0(t) = \dfrac{\sigma^2 m}{f^3}(e^{\alpha t} - 1 - \alpha t)$.

4 Yes. $X(t) = -\dfrac{1}{a_0}\displaystyle\int_t^\infty e^{\alpha(t-s)}\, dW(s)$.

5 (a) A solution to the indicated equation on an interval containing the point t_0 is defined as a process having continuous sample functions which satisfies the equation

$$a_0(X(t) - X(t_0)) + a_1 \int_{t_0}^{t} X(s)\, ds = c(t - t_0) + W(t) - W(t_0)$$

on that interval.

(c) $X_0(t) = -\dfrac{c}{a_0 \alpha} + \dfrac{1}{a_0} \int_{-\infty}^{t} e^{\alpha(t-s)}\, dW(s).$

6 (a) $\text{Var } X(t) = \sigma^2 \left(2e^{-t} - \dfrac{e^{-2t}}{2} - \dfrac{3}{2} + t \right).$

(b) $\text{Var } X(t) = \dfrac{\sigma^2}{12}(1 - 6e^{-2t} + 8e^{-3t} - 3e^{-4t}), \quad r_{X_0}(t) = \dfrac{\sigma^2}{12}(2e^{-|t|} - e^{-2|t|}).$

(c) $\text{Var } X(t) = \dfrac{\sigma^2}{80}[1 + e^{-2t}(4\cos t - 2\sin t - 5)],$

$r_{X_0}(t) = \dfrac{\sigma^2}{80} e^{-|t|} \left(\cos \dfrac{t}{2} + 2 \sin \dfrac{|t|}{2} \right).$

(d) $\text{Var } X(t) = \dfrac{\sigma^2}{4}[1 - e^{-2t}(2t^2 + 2t + 1)], \quad r_{X_0}(t) = \dfrac{\sigma^2}{4} e^{-|t|}(1 + |t|).$

(e) $\text{Var } X(t) = \dfrac{\sigma^2}{4}(2t - \sin 2t).$

9 (b) $\phi_1''(0) = -a_2/a_0, \quad \text{Var } X_0'(t) = \sigma^2/2a_0 a_1.$

11 (a) $\mu_{X_0}(t) = 1, \quad r_{X_0}(t) = e^{-|t|}(|t| + 1)/2.$

(b) $\mu_X(t) = 1 - e^{-t}, \quad r_X(s, t) = \dfrac{e^{-|t-s|}}{2}[|t - s| + 1 - (s + t + 1)e^{-2\min(s,t)}].$

13 (a) $\dfrac{r(1)}{r(0)} X(0), \quad \dfrac{r^2(0) - r^2(1)}{r(0)}.$

(b) $\dfrac{r(1)}{r(0)} X(0) - \dfrac{r'(1)}{r''(0)} X'(0), \quad \dfrac{r^2(0) - r^2(1)}{r(0)} - \dfrac{3(r'(1))^2}{r''(0)}.$

(c) $\left[\int_0^1 r(t)\, dt/(r(0) + r(1)) \right] (X(0) + X(1)),$

$\int_0^1 \int_0^1 r(s - t)\, ds\, dt - \dfrac{2(\int_0^1 r(t)\, dt)^2}{r(0) + r(1)}.$

16 $\hat{X}(t_1 + \tau) = e^{\alpha \tau} X(t_1), \quad$ where $\quad \alpha = -a_1/a_0; \quad (1 - e^{2\alpha\tau})/2a_0 a_1.$

17 (a) $X(t_1) + (1 - e^{-\tau})X'(t_1).$
(b) $(2e^{-\tau} - e^{-2\tau})X(t_1) + (e^{-\tau} - e^{-2\tau})X'(t_1).$

(c) $e^{-\tau} \left(\cos \dfrac{\tau}{2} + 2 \sin \dfrac{\tau}{2} \right) X(t_1) + 2e^{-\tau} \sin \dfrac{\tau}{2} X'(t_1).$

(d) $e^{-\tau}(1 + \tau)X(t_1) + \tau e^{-\tau} X'(t_1).$
(e) $\cos \tau\, X(t_1) + \sin \tau\, X'(t_1).$

Answers

19 (a) $h(t) = 1$, $\quad -\tfrac{1}{2} < t < \tfrac{1}{2}$, $\quad$ and $\quad h(t) = 0$ elsewhere.

(b) $f_X(\lambda) = \left(\dfrac{\sin \lambda/2}{\lambda/2}\right)^2 f_Y(\lambda)$.

22 $\dfrac{\sigma^2}{\pi}\left(\dfrac{1 - \cos \lambda}{\lambda^2}\right)$.

23 $1/\pi(1 + \lambda^2)^2$.

26 $F(\lambda) = \begin{cases} 0, & -\infty < \lambda < -\lambda_2, \\ \sigma_2^2/2, & -\lambda_2 \leq \lambda < -\lambda_1, \\ (\sigma_1^2 + \sigma_2^2)/2, & -\lambda_1 \leq \lambda < \lambda_1, \\ \sigma_1^2 + (\sigma_2^2/2), & \lambda_1 \leq \lambda < \lambda_2, \\ \sigma_1^2 + \sigma_2^2, & \lambda_2 \leq \lambda. \end{cases}$

Glossary of Notation

CHAPTER 1

$\mathscr{S}$	state space
X_n	state of system at time n
π_0	initial distribution
$P(x, y)$	probability of going from x to y in one step
$P^n(x, y)$	probability of going from x to y in n steps
$P_x(\ )$	probability of an event defined in terms of a chain starting at x
T_A	hitting time of the set A
T_y	hitting time of the state y
ρ_{xy}	probability that a chain starting at x will ever visit y
$1_y(x)$	function that is one if $x = y$ and zero if $x \neq y$
$N(y)$	total number of visits to y
$E_x(\ )$	expectation of a random variable defined in terms of a chain starting at x
$G(x, y)$	expected number of visits to y for a chain starting at x
$\mathscr{S}_T$	set of transient states
$\mathscr{S}_R$	set of recurrent states
$\rho_C(x)$	probability that a chain starting at x will eventually be absorbed into the closed set C

CHAPTER 2

π	stationary distribution
$N_n(y)$	number of visits to y by time n
$G_n(x, y)$	expected number of visits to y by time n for a chain starting at x
m_y	mean return time to a recurrent state y
T_y^r	time of rth visit to y
W_y^r	waiting time between the $(r-1)$th visit to y and the rth visit to y
$\mathscr{S}_P$	set of positive recurrent states
d_x	period of the state x
d	period of an irreducible chain

CHAPTER 3

$\mathscr{S}$	state space
τ_n	time of nth jump
$X(t)$	state of system at time t
$P_x(\ )$	probability of an event defined in terms of a process starting at x
$E_x(\ )$	expectation of a random variable defined in terms of a process starting at x
$F_x(t)$	distribution function of time to first jump for a process starting at a non-absorbing state x
δ_{xy}	one if $x = y$ and zero if $x \neq y$
Q_{xy}	probability that a process starting at a non-absorbing state x will go to y at its first transition ($Q_{xy} = \delta_{xy}$ if x is an absorbing state)
$P_{xy}(t)$	probability that a process starting at x will be in state y at time t
π_0	initial distribution of the process
q_x	exponential parameter in distribution of time to first jump for a process starting at a non-absorbing state x ($q_x = 0$ if x is an absorbing state)
q_{xy}	infinitesimal parameters of the process defined by $q_{xy} = P'_{xy}(0)$

T_y	first time $t \geq \tau_1$ that the process is in state y	$r_X(t)$	covariance between $X(s)$ and $X(s+t)$ for a second order stationary process
ρ_{xy}	δ_{xy} if x is an absorbing state; otherwise the probability that the process visits y at some time $t \geq \tau_1$	$r_{XY}(s,t)$	covariance between $X(s)$ and $Y(t)$
π	stationary distribution of the process		
m_x	mean return time to a non-absorbing recurrent state x		

CHAPTER 5

$X'(t)$ derivative of the $X(t)$ process

$X^{(n)}(t)$ nth derivative of the $X(t)$ process

CHAPTER 4

T	time parameter set
$X(t)$	value of process at time t
$\mu_X(t)$	mean of $X(t)$
$r_X(s,t)$	covariance between $X(s)$ and $X(t)$
μ_X	mean when independent of t

CHAPTER 6

$\mathscr{M}$	collection of allowable estimators
$\hat{Y}$	estimator of Y
f_X	spectral density function
F_X	spectral distribution function

Index

Absorbing state, chain, 8
   process, 85
Absorption probabilities, 25
Allowable estimators, 171
Aperiodic chain, 73
Aperiodicity of a process, 104
Auto-covariance function, 111
   second order stationary process, 113

Backward equation, 89
   birth and death process, 92
Birth and death chain, 9
   period, 73
   positive recurrence, 66
   recurrence, 32
   stationary distribution, 50
Birth and death process, 89
   stationary distribution, 104
   transience, null recurrence, and positive recurrence, 104
Birth rates, 89
Bochner's theorem, 182
Bounded convergence theorem, 63
Branching chain, 10
   extinction probability, 11, 34
Branching process, 91
   immigration, 97
Brownian motion, 123
   nondifferentiability, 141

Chapman-Kolmogorov equation, 87
Characteristic polynomial, 160
Closed set of states, 22
   irreducible, 23
Continuous parameter process, 111

Covariance function, 111
   auto-, 111
   continuous, 128
   cross-, 118
   nonnegative definite property, 112
   second order stationary process, 113
   symmetry, 112
Cross-covariance function, 118
   continuity, 129

Death rates, 89
Differentiable second order process, 135
Discrete parameter process, 111
Distribution of time to first jump, 85
   exponential, 86
Divisor, 72
Doubly stochastic transition function, 82

Ehrenfest chain, 7
   modified, 52
Einstein, Albert, 123
Embedded chain, 102
Empirical distribution function, 125
Estimation, 170
   general principle, 173
   linear, 172
   mean square error, 171
   nonlinear, 172
   optimal, 171
Estimation of mean, 149
Estimation of signal, 188
Expected number of visits, 19
   by time n, 57
Explosions, 85
Exponential parameter, 87

Filtering, 171
Finite linear combination, 112
Forward equation, 89
   birth and death process, 92
Frequency response function, 180

Gambler's ruin chain, 8
Gaussian distribution, 120
   joint, 121
Gaussian process, 119
   differentiation, 139
   integration, 134
   sample function continuity, 131
   strictly stationary, 122
Greatest common divisor, 72

Hitting time, chain, 14
   process, 102

Impulse response function, 160
Indicator function, 18
Infinite server queue, 99
   positive recurrence, 106
Infinitesimal parameters, 89
Initial distribution, chain, 5, 6
   process, 86
Irreducible chain, 23
Irreducible closed set, 23
Irreducible process, 102

Jump process, 84
   explosive, 86
   non-explosive, 85
   pure, 85
Jump times, 84

Langevin's velocity process, 184
Lévy, Paul, 123
Linear birth process, 98
Linear estimation, 172

Markov chain, 1
   aperiodic, 73
   irreducible, 23
   null recurrent, 62
   periodic, 73
   positive recurrent, 62
   recurrent, 21
   transient, 21
Markov property, chain, 1
   process, 86

Markov pure jump process, 86
   aperiodicity, 104
   irreducible, 102
   null recurrent, 103
   positive recurrent, 103
   recurrent, 102
   transient, 102
Martingale, 27
Mean function, 111
   continuity, 128
Mean return time, chain, 58
   process, 103
Mean square continuity, 128
Mean square convergence, 138, 172
Mean square error, 171

N server queue, 106
Nonlinear estimation, 172
Nonnegative definite property, 112
Normal distribution, joint, 121
Normal process, 120
Null recurrent chain, 62
Null recurrent process, 103
Null recurrent state, chain, 60
   process, 103
Number of visits, 18
   by time n, 57

One-step transition probabilities, 5
Optimal estimator, 171
Ornstein-Uhlenbeck process, 184
Orthogonal random variables, 173

Period of a chain, 73
Period of a state, 72
Periodic chain, 73
Poisson process, 96
   conditional distribution of arrivals, 99
   mean and covariance functions, 115
Positive recurrent chain, 62
Positive recurrent process, 103
Positive recurrent state, chain, 61
   process, 103
Prediction, 170
Probability generating function, 34
Pure birth process, 90
   linear, 98
Pure death process, 90

Index

Queuing chain, 9
    positive recurrence, 69
    recurrence, 36
Queuing process, infinite server, 99
    N server, 106

Random walk, 7
    simple, 7
Recurrent chain, 21
Recurrent process, 102
Recurrent state, chain, 17
    process, 102

Sample function continuity, 130
Schwarz's inequality, 113
Second order process, 111
    derivative, 135
    higher derivatives, 137
    stationary, 112
Second order stationary process, 112
Spectral density function, 177, 183
Spectral distribution function, 178, 182
Stability, 160
State space, chain, 1
    process, 84
Stationary distribution, chain, 47
    concentrated on a closed set, 67
    process, 102
Stationary transition probabilities, 1
Steady state distribution, 47
Stochastic differential equation, 152
    nth order, 153, 165, 166
Stochastic process, 111
Strictly stationary process, 122
Strong law of large numbers, 58

Time parameter set, 111
Transient chain, 21
Transient process, 102
Transient state, chain, 17
    process, 102
Transition function, chain, 5, 6
    m-step, 13
    process, 86
Transition matrix, 16
    n-step, 16
Transition probabilities, chain, 1
    one-step, 5
    process, 85
    stationary, 1
Two-state birth and death process, 92
    mean and covariance function, 114
Two-state Markov chain, 2, 17, 49

White noise, 142
    spectral density, 182
Wiener, Norbert, 123
Wiener process, 123
    continuity of paths, 132
    covariance function, 124
    non-differentiability, 140
    white noise, 142
Wiener-Lévy process, 123